Luciana Godinho

Scenedesmaceae family in the state of São Paulo Floristic survey

AF294451

Luciana Godinho

Scenedesmaceae family in the state of São Paulo Floristic survey

Chlorococcales - Clorofíceas

Imprint
Any brand names and product names mentioned in this book are subject to trademark, brand or patent protection and are trademarks or registered trademarks of their respective holders. The use of brand names, product names, common names, trade names, product descriptions etc. even without a particular marking in this work is in no way to be construed to mean that such names may be regarded as unrestricted in respect of trademark and brand protection legislation and could thus be used by anyone.

Cover image: www.ingimage.com

This book is a translation from the original published under ISBN 978-613-9-61270-3.

Publisher:
Sciencia Scripts
is a trademark of
Dodo Books Indian Ocean Ltd. and OmniScriptum S.R.L publishing group

120 High Road, East Finchley, London, N2 9ED, United Kingdom
Str. Armeneasca 28/1, office 1, Chisinau MD-2012, Republic of Moldova, Europe
Printed at: see last page
ISBN: 978-620-7-62150-7

Table of contents:

LUCIANA RUFINO GODINHO

Family Scenedesmaceae (Chlorococcales, Chlorophyceae) in the State of São Paulo: floristic survey.

ACKNOWLEDGMENTS

During the course of this work, many people followed my steps and were therefore directly or indirectly important and even responsible for its realization. These people are the following and I would like to express to them a small token of my immense affection and sincere gratitude:

To my advisor, Prof. Dr. **Carlos Eduardo de Mattos Bicudo**, Scientific Researcher at the Ecology Section of the Botanical Institute of the São Paulo State Department of the Environment, for his wisdom, patience, understanding and for being a great person to work with over the years. Thank you so much for being the wonderful person you have always been and for being a true scientific father who contributed intensely to the development of this work. Thank you from the bottom of my heart for never letting me give up and for always supporting me in good times and, above all, in difficult times. Thank you for helping me grow as a professional and as a person and for making me believe that we should never give up on our dreams, no matter how far away they may seem. I would like to say that you have always been and will always be more than a scientific father and that it is an honor to be your mentor. Ah, if only everyone in the world were like you!

To Prof.[3] Dr.[a] **Denise de Campos Bicudo**, Scientific Researcher at the Ecology Department of the Botanical Institute of the São Paulo State Department of the Environment, for her friendship, support, belief in me and my professional abilities and for her ever-present professional example. An admirable professional and a charming person. I always say that when I grow up, I want to be just like her.

To Prof. Dr. **Augusto A. Comas G.**, Scientific Researcher at the Center for Environmental Studies in Cienfuegos, Cuba. I would like to thank you for the fantastic experience in Cuba. For reviewing and confirming the results of our work. For being such a wonderful person who, spiritually speaking, is even better. For all the support, patience, dedication and care you gave to this thesis until the end. I don't have many words to sum up how grateful I am for everything the professor has done and has done. I would just like to say that you have a special place in my heart. Thank you so much!

To Prof.[3] Dr.[a] **Carla Ferragut**, Scientific Researcher at the Ecology Section of the Botany Institute of the São Paulo State Department of the Environment, for being such a special person who possesses so much knowledge. I often say that you are a little piece of the 'Barsa' encyclopedia and that you never think twice before sharing your knowledge and professional experience with those who are and those who are not your students. Thank you for proofreading part of this work and for always being such a companion and so helpful.

To all the **researchers in** the Ecology Section, for their support and pleasant interaction.

To our **friends** Angélica, Bárbara, Camila (for their hard work on the boards), Cati, Fernanda, Carla Ferragut, Joy, Kleber (Phycology), Lu Fontana, Lu Mineira, Murilo, Norminha, Thiago, Samanta, Sandrinha and Yukio (for all their help); to those who have finished their postgraduate programs and no longer attend the Ecology Section: Dani and Sidney; and newcomers Jeniffer and Simone; for their conviviality, support and good laughs. As well as the terrestrial staff who are always there.

To the research support staff of the Ecology Section, **Amariles, Dorinha, Marli and Val**, for their immense collaboration, pleasant socializing and the parties that were always full of delights.

To the Postgraduate Coordination of the Botanical Institute, for their support and confidence in this work.

To the management of the Botanical Institute of the São Paulo State Secretariat for the Environment, for giving me unrestricted use of all its facilities and infrastructure during all these years of work.

To FAPESP, the São Paulo State Research Foundation, for its financial support through the BIOTA/FAPESP Program, the Virtual Biodiversity Institute, and the granting of a doctoral scholarship (Process No.[0] **06/53317-8**), which were essential for the materialization of this work.

To my everyday **friends** and sisters **Camilinha**, the **J** trio and **Miloca, Vanessinha, Licia, Lelé, Sarita, Bruninha** and **Jacozito**, for supporting me, giving me so much strength, for all the fantastic places we've been together and for the affection and friendship that is what keeps us together, side by side always!

To all the **friends** and **family** who have also participated in some way and encouraged my work. Thank you all.

To my godmother **Marineves**, for being an example of life, and to my cousin **Paulo Gazzani Júnior**, my great professional and personal inspiration, for all the support, affection and for always believing in my professional ability, as well as for being an intense part of my life and my future. I love you!

And to my parents **Walter** and **Marilene** and my brother **Leonardo**, for so often giving up their time to look after Pepé because of my work; for giving me so much strength full-time; for always being by my side; and for showing me that only by going to war and being humble will we achieve anything in our lives. But most of all, for showing me that the most important things in life are **FAMILY** and **LOVE.**

And to my little son, **Pedro Henrique**, for being such a companion and partner at all times and for understanding that having a new mother is neither common nor easy. Thank you, my little angel, for everything! You are the one and only true owner of my heart, the most beautiful face of this new generation. I LOVE YOU!

Summary

Knowledge of the diversity of the Scenedesmaceae family (Chlorocaccales, Chlorophyceae) in the state of Sao Paulo, Brazil was, until then, relatively small, including in terms of its geographical distribution.

The current floristic survey of representatives of the family in the state of São Paulo is based on the study of 153 sample units deposited in the State Scientific Herbarium "Maria Eneyda P. Kauffmann Fidalgo" (SP) of the Botanical Institute. Materials from 122 municipalities were examined and specimens of Scenedesmaceae were documented in 91 of them.

73 Scenedesmaceae taxa were identified, distributed in 66 species (including three possible species new to science), 12 varieties that are not typical of their respective species and two new combinations. Of the 73 taxa identified, nine were cited for the first time for the state of Sao Paulo. At the same time, 22 taxa in the specialized literature were evaluated, which were not found in the samples studied, but presented a description, measurements and illustration of the material studied, i.e. sufficient conditions for their re-identification. In addition to these 22 taxa, there is a list of 75 others which have currently been excluded because they are not suitable for study and re-identification. Finally, 46 other taxa that had been wrongly identified and therefore appear under different names in the literature were maintained.

The identifications were made, whenever possible, from the study of populations, based on the classic morphological, meristic and metric characteristics of the cells, in order to assess intra- and inter-population variation.

For each taxon identified, the following information was provided: (/) the complete bibliographic reference of the work containing its original description and/or diagnosis; (2) a detailed morphological description including measurements of taxonomic interest; (3) a list of the samples in which the taxon was found; (4) geographical distribution in the state of Sao Paulo; (5) taxonomic comments; and (6) illustrations. Where available, the basionymes have also been indicated.

The most widely distributed taxa in the area of the State of São Paulo, considering the 122 municipalities analyzed, were: *Scenedesmus obliquus* (Turpin) Kützing var. *dimorphus (*Turpin) Hansgirg (in 22 municipalities), *Desmodesmus brasiliensis* (Bohlin) Hegewald (in 17 municipalities) and *D. armatus (*Chodat) Hegewald var. armatus, *D. communis (*Turpin) Hegewald and .S' acunae Comas (in 14 municipalities each).

The taxa in which the greatest occurrence of polymorphism was found were the following: *Desmodesmus brasiliensis* (Bohlin) Hegewald, *D. armatus* (Chodat) Hegewald var. *armatus, D. abundans* (Kirchner) Chodat, *D. arthrodesmiformis* (Schroder) An, Friedl & Hegew and S. *obliquus* (Turpin) Kützing var. *dimorphus (Turpin)* Hansgirg.

The identifications of *Desmodesmus* sp. 1, *Desmodesmus sp.* 2 and *Desmodesmus* sp. 3 and the new combinations of *Scenedesmus aculeolatus* Reinsch and *Scenedesmus gutwinskii* Chodat var. *bekesensis* Uherkovich are practically definitive and will have their formal proposition pending the analysis of a greater number of specimens to be sure of the stability, at population level, of the characteristics currently considered diagnostic.

Finally, the importance of population analysis for the taxonomic identification of species and taxonomic varieties of Scenedesmaceae was highlighted due to the frequent occurrence of polymorphism and the need to identify the morphological characteristics that are the most reliable for delimiting and defining the different taxonomic categories of the family. The morphological characteristics of the vegetative life that showed the greatest stability in the populations currently examined were: cell shape, number of cells per cenobium, which is generally four, and the presence or absence of pyrenoids. These characteristics should therefore be used to separate species and varieties, depending on the level of variation presented. The morphological characteristics of the vegetative life that showed the greatest variability in the populations examined were: shape of the margin of the outer cells of the cenobium, size and orientation of the polar spines and measurements of cell length and width. These characteristics should therefore not be used to separate species and varieties in Scenedesmaceae. At most, if the stability of these latter characteristics is significant enough, they can be used to separate taxonomic forms.

Chapter 1
Introducing

1.1. Ordern Chlorococcales *'sensu lato''*

Unicellular green algae with a coccal level of vegetative organization (FOTT 1971), immobile in the vegetative state, in which some species may present mobile phases by flagella during reproduction, are classified in the Class Chlorophyceae. MARCHAND (1895) defined the Order Chlorococcales for this group of algae, but the name Protococcales Wille 1897 was more commonly used until the exclusion of the genus *Protococcus* C. Agardh from the order (PASCHER 1915). It wasn't until WEST & FRITSCH (1927) that the term Chlorococcales was fully accepted.

Among the morphological attributes that define the Order Chlorococcales are the following: unicellular, solitary or colonial green algae, whose asexual reproduction occurs by means of aplanospores, zoospores, hemizospores or autospores, while sexual processes (iso, aniso or oogamy) are known in a few genera (KOMÁREK & FOTT 1983).

The taxonomic composition of the Order Chlorococcales has undergone important changes since it was first proposed (WILLE 1897, 1909, BRUNNTHALER 1915, PRINTZ 1927, KORSIKOV 1953, BOURRELLY 1990, KOMÁREK & FOTT 1983).

There are four monographs which, in one way or another, bring together all the information on the order disseminated in the literature. Under the name Protococcales, there is the work by BRUNNTHALER (1915) published within the series organized by Adolf Pascher entitled "Süsswaßerflora von Deutschland, Österreich und der Schweiz", which divides the order into two suborders depending on the type of their asexual reproduction, whether by zoospores (Suborder Zoosporinae) or by autospores (Suborder Autosporinae) and recognizes nine families within it. Another monographic work is that of KORSIKOV (1953), published as part of the flora of Ukraine and establishing the following two orders according to the presence or absence of contractile vacuoles, respectively Vacuolales and Protococcales proper. Under the name Chlorococcales, PHILIPOSE (1967) published, in a monograph on the Chlorococcales of India, a historical review of all the changes it had undergone, defining the taxonomy of the order. Finally, KOMÁREK & FOTT (1983) published the first worldwide monograph on the Chlorococcales, based on traditional criteria but including more recent scientific contributions. In this work, the authors maintained the fundamental concept of the Order Chlorococcales defined by WILLE (1909), BRUNNTHLER (1915), PRINTZ (1927) and BOURRELLY (1990). In the system they adopted, these authors included, albeit provisionally, the suborders Spongiococcoideae, Chlorosarcinoideae and Glaucocystoideae. The heterogeneity of the group and the possible identification of several evolutionary lines within it were considered by KOMÁREK & FOTT (1983), who took into account, fundamentally, cytological characteristics (uni- or plurinucleated cells), ultrastructural features of the cell wall and the type of reproduction.

There are important monographs for some genera of Chlorococcales or for certain systematic groups, such as those by REHÁKOVA (1969) on *Oocystis;* by KOMÁRKOVÁ- LEGNEROVÁ (1969) on *Ankistrodesmus* and *Monoraphidium;* KOMÁREK (1974) on the Crucigenioideae; Frantisek Hindák's important contribution spread over several works (HINDÁK 1977, 1980, 1984, 1988, 1990); and the works by PHILIPOSE (1967), KOMÁREK (1983) and COMAS (1996).

ETTL (1980) and ETTL & KOMÁREK (1982) considered that in the Family *Chlorococcaceae,* whose genus-type is *Chlorococcum,* reproduction occurs by zoospores of the *Chlamydomonas* type, which represent a well-defined and relatively narrow evolutionary line, which should be separated from the group as an independent order, within a special class, the Chlamydophyceae. All the other genera of Chlorococcales *'sensu lato'* whose reproduction is exclusively by autospores were included by BOLD & WYNNE (1978) in the Order Chlorellales.

Advances in electron microscopy and biochemical, physiological and ecological studies have shown that the Chlorococcales are not a monophyletic group, but have several lines with their own historical evolution (ETTL 1981, ROGERS *ET AL.* 1980, ETTL & KOMÁREK 1982, MATTOX & STEWART 1984, KOMÁREK 1987), establishing the following orders: Chlorococcales *'sensu stricto''* (reproduction by zoospores of the *Chlamydomonas* type*),* Protosiphonales (multinucleated cells and reproduction by zoospores of the *Dunaliella* type*),* Chlorellales (reproduction exclusively by autospores) and Chlorokybales (sarcinoid habit and reproduction by zoospores with flagella similar to those of the Charophyceae).

The ultrastructure of the mitotic processes and of the components of the flagellar apparatus (kinetosome) reflects on phylogenetic relationships and has served as a basis for the modern systematics of green algae (MATTOX & STEWART 1984, O'KELLY & FLOYD 1984, MELKONIAN 1990), fostering arguments for the existence of polyphyletic lines within this group (WATANABE & FLOTO 1992, FLOTO *ETAL.* 1993). According to these criteria, the following four types of flagellar apparatus stand out within the Chlorococcales *'sensu lato":* (7) directly opposite (DO) present in the families Hydrodictyaceae and Neochloridaceae; (2) clockwise (CW) present in the family Chlorococcaceae *'pro parte', (3)* counterclockwise (CCW) present in *Trebouxia,* a genus formerly classified among the Chlorococcaceae and, more recently, in the Class Trebouxiophyceae (FRIEDL 1995); and (4) the group represented by *Chlorokybus,* with other flagellar characteristics that justified its transfer to the Chlorokybales (Streptophytina).

Currently included in the Order Sphaeropleales (DEASON *ETAL.* 1991) are unicellular, filamentous or cenobial chlorophytes that produce, in the reproductive phase, mobile biflagellate cells with directly opposed basal bodies (DO).

Molecular genetic studies supported certain changes established by cytological and ultrastructural research (MELKONIAN & SUREK 1995, FRIEDL 1995). Since then, there has been a veritable revolution in the systematics of the Chlorococcales, based on the study of ribosomal DNA sequences, especially the 18S fraction, which later included the large subunit (26S) (FRIEDL & ROKITTA 1997, BUCHHEIM *ET AL.* 2001) and the ITS-2 spacer, in order to establish phylogeny at genus and species level (AN *ET AL.* 1999, KRIENITZ *ET AL.* 2004). These studies reached some unexpected conclusions, such as the phylogenetic relationships between the Trebouxiophyceae (precisely defined by the CCW orientation of the flagella) and the species of *Chlorella* and *Oocystis,* which have neither flagella nor traces of kinetosome at any stage of their life history. Something similar occurred with the Sphaeropleales *'sensu'* DEASON *ET AL.* (1991) and their phylogenetic relationships with various genera completely devoid of flagella, such as *Coelastrum, Scenedesmus, Desmodesmus* and others. Sphaeropleales are characterized by directly oppositely oriented (DO) flagella.

In a synoptic way, COMAS & SÁNCHEZ (2008) summarized the implications of molecular genetic studies in the taxonomy and systematics of unicellular green algae as follows:

(1) **Antagonism between traditional taxonomy (morphology) and genetic analysis**. Many characters considered important do not play a role in the differentiation of some taxa, nor in phylogenetic relationships, such as the Selenastraceae Family, a relevant case that shows the close relationships between morphologically very different species such as *Closteriopsis acicularis* (G.M Smith) Belcher & Swale and *Parachlorella kessleri* (Fott & Nováková) Krienitz.

(2) **Relevant systematic changes**. The families Oocystaceae, Chlorellaceae *"pro parte"* and Botryococcaceae *"pro parte"* have been transferred to the Class Trebouxiophyceae. The families Scenedesmaceae (which includes the Family Coelastraceae *"pro parte"),* Hydrodictyaceae, Neochloridaceae and Selenastraceae *"pro parte"* were transferred to the Order Sphaeropleales *"sensu"* DEASONETAL. (1991).

(3) **Concept of variable taxa**. This includes genera taken in their broad sense such as *Ankistrodesmus* (including *Monoraphidium, Podohedriella* and *Quadrigula)* and *Kirchneriella* (including *Selenastrum)* and genera taken in their narrow sense such as *Pediastrum* Meyen, which has recently been divided into five genera *(Pediastrum "sensu stricto", Pseudopediastrum* Hegewald, *Parapediastrum* Hegewald, *Stauridium* Corda and *Monactinus* Corda.

1.2. Family Scenedesmaceae Oltmanns 1904

The system used in this work was that of KOMÁREK & FOTT (1983), i.e. the classic taxonomic system was chosen.

The Scenedesmaceae comprise 30 genera and around 450 species, of which the most specious and representative genus is *Scenedesmus.* Also belonging to this family, in alphabetical order, are the genera: *Coronastrum, Crucigenia, Crucigeniella, Danubia, Desmodesmus, Dicloster, Didymogenes, Dimorphococcus, Enallax, Gilbertsmithia, Komarekia, Lauterborniella, Makinoella, Neodesmus, Pseudodidymocystis, Pseudotetradesmus, Pseudotetrastrum, Rayssiella, Scenedesmus, Schmidleia, Schroederiella, Suxeniella, Tetrachlorella, Tetradesmus, Tetrallantos, Tetranephris, Tetrastrum, Westella, Westellopsis* and *Willea.*

Traditionally, the Scenedesmaceae Family is classified among the green algae, including unicellular forms capable of forming cenobia more or less aligned in one or more planes and basically made up of four cells. The arrangement of the cells and the way in which they are joined to each other, i.e. directly by their cell walls or by projections from the wall itself, have served to define subfamilies (KOMÁREK 1974, KOMÁREK & FOTT 1983) as follows: *(P)* Danubioideae ^ cenobia with four cells touching at one end; the cells are more or less regular and can be joined by remnants of the maternal walls to form syncenobia; (2) Coronastroideae ^ cells joined by delicate gelatinous appendages at one or both poles, forming more or less square cenobia, whose longitudinal axes of the cells are perpendicular or even slightly inclined in relation to the plane of the cenobium; they can sometimes form syncenobia; (3) Crucigenioideae ^ flat cenobiums formed by four cells arranged in a cross, forming two pairs, one above the other, or by two inner cells more or less parallel to each other and two outer cells joined obliquely to the previous pair by their poles; (4) Tetrallantoideae ^ cenobiums of various types consisting of two groups of four cells, each forming an angle of 45° to the other, by two annular groups of four cells, each cell of each group joining at its ends to form a crown-shaped cenobium; or by cenobiums of four cells, two of which are oriented more or less parallel to each other, while the other two each rest on the point where the first pair of cells join; (5) Scenedesmoideae ^ linear, flat cenobiums made up of one or two rows of cells, rarely semicircular, with the longitudinal axis of the cells perpendicular to each other or slightly inclined in relation to the longitudinal axis of the cenobium, sometimes with cells joined by their convex margins; and *(6)* Dimorphococcoideae ^ cenobia with 4, 8 or 16 cells markedly alternating with each other, arranged in different planes of the spathe, sometimes with morphological differences between the outer and inner cells and often with the daughter cenobia joined by remnants of the maternal walls to form syncenobia.

The cell wall of representatives of the Scenedesmaceae Family is made up of an inner layer of cellulose and one or more layers of sporopollenin (ATKINSON *ETAL.* 1972), the outermost of which may have ornaments or various structures.

Reproduction takes place by autospores united in autocoenobes inside the mother cell. There are records of motile phases in *Scenedesmus obliquus* (Turpin) Kützing (TRAINOR 1963, 1965, CAIN & TRAINOR 1976, LUKAVSKY 1991).

TRAINOR (1963, 1965) found mobile stages in this species which they identified as zoospores, but CAIN & TRAINOR (1976) considered them to be gametes. However, the latter authors did not observe the germination of zygotes. LUKAVSKY (1991) found mobile cells and observed how a flagellated cell transformed into a vegetative cell, which were therefore zoospores. The statement "mobile phases in *S. obliquus"* means that there is still no definition of whether they are zoospores or gametes.

TRAINOR *ETAL.* (1991) provided examples of morphological variation in microalgae, including diatoms, chlorophyceae and cyanobacteria, and their possible implications for the taxonomy of some species, commenting that the number and position of spines in various strains of *Scenedesmus* are not stable. They also mentioned that cultures of *Lagerheimia* have produced several colonies that are morphologically identical to those of *Scenedesmus*.

Molecular studies have included *Scenedesmus* in the Order Sphaeropleales *'sensu* DEASON *ET AL.* (1991), which includes *Coelastrum* Nageli and therefore the Family Coelastraceae WILLE (1909) according to KRIENITZ *ET AL.* (2003). This order originally included only filamentous and multinucleate species with very particular sexual reproduction. However, DEASON *etAL.* (1991) also included unicellular, uninucleate species without sexual reproduction. These authors showed that *Coelastrum* should be classified in the Family Scenedesmaceae. Also according to the same criteria, some genera and species have been removed or excluded from the Family Scenedesmaceae, such as *Dicloster* Jao (Trebouxiophyceae) (HEGEWALD & HANAGATA 2000), *Didymocystis, Crucigeniella rectangularis* (Nageli) Komárek and *Tetrachlorella alternans* (G.M. Smith) Korsikov (Oocystaceae, Trebouxiophyceae) (HEGEWALD 1988, KRIENITZ *ET AL.* 2003, HEGEWALD & HANAGATA 2000).

JOHNSON *ET al.* (2007) stated that both *Scenedesmus* and *Desmodesmus* species are extremely common in continental waters worldwide. They further stated that the enormous phenotypic plasticity commonly demonstrated by these algae, together with the minute differences between different species of each genus, makes taxonomic identification of these materials extremely difficult under light microscopy, especially when working directly with material collected from the environment. JOHNSON ET AL. (2007) worked with cultures of material from Itasca State Park located in Minnesota, USA. From the 100 cultures examined, they isolated 34 types of sequences using the rDNA ITS-2 spacer. Twenty-four of these sequences were sufficiently distinct from each other to correspond to distinct taxonomic units. However, these results exceeded what optical microscopy could identify in previous studies carried out with material from the Park. Five of the isolated sequences matched species identified by light microscopy and the sequences of six other isolates did not. JOHNSON *ET AL.* (2007) concluded that cultivation and genetic sequencing using the rDNA ITS-2 spreader is an efficient tool for identifying the specific diversity level of *Scenedesmus* and *Desmodesmus*.

1.3. Family Scenedesmaceae in Brazil

The literature on the Family Scenedesmaceae in Brazil shows that few authors have developed specific studies on this family and that today there is still a great lack of publications on the Chlorococcales. However, there are some very important Brazilian works which contain descriptions and illustrations and, in some cases, also keys for identifying the material studied.

In chronological order, SANT'ANNA & MARTINS (1982) studied the Chlorococcales of lakes Cristalino and Sao Sebastiao, both located in the state of Amazonas; PICELLI-VICENTIM (1987) studied the planktonic Chlorococcales of Iguacu Park in Curitiba, Paraná State; ROLLASE AL. (1990) studied the limnological aspects of the Volta Grande Reservoir, in the state of Minas Gerais, including the Chlorococcales; ROSA & OLIVEIRA (1990) worked with the Chlorococcales of bodies of water in the municipality of Sao Jerónimo, in the state of Rio Grande do Sul; NOGUEIRA (1991) inventoried and commented in depth on the Chlorococcales of the Municipality of Rio de Janeiro and its surroundings; MARTINS-DASILVA (1996) studied the new occurrences of Chlorophyceae for the State of Pará; and BICUDO & MENEZES (2006) published a key for identifying genera of algae from Brazilian continental waters. The latter work contains descriptions and illustrations, as well as recommending literature for infrageneric identification.

1.4. Family Scenedesmaceae in the State of São Paulo

The first work to record the occurrence of Scenedesmaceae material in the state of Sao Paulo was by WITTROCK & NORDSTEDT (1880). It is a collection of exsiccates of algae collected in freshwater and marine environments, gathered in fascicles of 50 exsiccates each, distributed by V.B. Wittrock and C.F.O. Nordstedt between 1877 and 1889 and by V.B. Wittrock, C.F.O. Nordstedt and E.G. Lagerheim between 1893 and 1903. Description and original illustration of *Scenedesmus acutus* Meyen are part, in the form of a label, of exsiccate no.^{0} 351 of fascicle no.^{0} 8 of the aforementioned collection. The material in this exsiccate was collected in a place near the farm of Dr. Martim Francisco Ribeiro de Andrada, in the area where the Campo de Aviacao de Marte is today, in the city of Sao Paulo.

The index prepared by EDWALL (1896) of the plants deposited in the old herbarium of the Geographical and Geological Commission of Sao Paulo contains *Scenedesmus acutus* Meyen, but without any description or illustration.

BORGE (1918) is the result of an examination of 239 samples collected by Alberto Lofgren in and around the city of Sao Paulo, where he lived, and in and around the city of Pirassununga. Eight species of Scenedesmaceae are mentioned in this work, along with various morphological expressions that were not formally proposed as taxonomic novelties. All the materials were documented with brief descriptions and few illustrations and are as follows: *Scenedesmus obliquus* (Turpin) Kützing, *Scenedesmus obliquus* (Turpin) Kützing var. *dimorphus* (Turpin) Rabenhorst, *Desmodesmus brasiliensis* (Bohlin) Hegewald (as *Scenedesmus brasiliensis* Bohlin), *Desmodesmus communis (Turpin)* Hegewald [as Scenedesmus *quadricauda (Turpin)* Brébisson], Scenedesmus *bifugatus (Turpin)* Kützing, *Scenedesmus bifugatus (Turpin)* Kützing var. *alternans* (Reinsch) Hansgirg, *Crucigenia rectangularis* (Nageli) Gay and *Dimorphococcus lunatus* A. Braun. All eight species were identified from material collected in the city of Pirassununga.

In 1948, Frederico Carlos Hoehne published the book "Plantas Aquáticas" (Aquatic Plants) and in it he cited and illustrated only one species of Scenedesmaceae, *Desmodesmus brasiliensis (*Bohlin) Hegewald (as *Scenedesmus brasiliensis* Bohlin) (HOEHNE 1948).

The work "Genera of freshwater algae from the city of Sao Paulo and its surroundings" published in 1963 by Aylthon Brandao Joly contains a description and illustration of the Chlorococcal genera *Dimorphococcus* A. Braun, *Scenedesmus* Meyen and *Tetrallantos* Teiling (JOLY 1963).

In 1977, Kozo Hino and José Galizia Tundisi published "Atlas das algas da Represa do Broa", which included descriptions of three genera - *Dimorphococcus* A. Braun, *Scenedesmus* Meyen and *Tetrallantos* Teiling - and illustrations of the following seven species: *Dimorphococcus lunatus* A. Braun, *Scenedesmus acutus* Meyen, Scenedesmus *acuminatus* (Lagerheim) Chodat, *Scenedesmus ecornis* (Ralfs) Chodat, *Desmodesmus microspina* (Chodat) Hegewald (as Scenedesmus *microspina* Chodat), *Desmodesmus communis (*Turpin) Hegewald [as *Scenedesmus quadricauda (*Turpin) Brébisson] and *Tetrallantos lagerheimii* Teiling (HINO & TUNDISI 1977).

LEITE & BICUDO (1977) proposed a new genus of Chlorococcales, *Tetranephris* Leite & C. Bicudo, from material collected in the south of São Paulo State, which they classified in the Family Scenedesmaceae represented by the single species *T. brasiliensis* Leite & C. Bicudo.

The nomenclatural evaluation of *Scenedesmus bifugus* (Turpin) Kützing (Chlorophyceae,

Scenedesmaceae) carried out in 1981 by Célia Leite Sant'Anna and Carlos Eduardo de Mattos Bicudo discussed the correct name of the species according to the International Code of Botanical Nomenclature (SANT'ANNA & BICUDO 1981).

There are 11 works of a purely taxonomic nature carried out with material from the state of São Paulo, which include representatives of the Scenedesmaceae family. They are: LEITE (1974, 1979), CARDOSO (1979), SANT'ANNA (1984), SANT'ANNA *ET AL. (*1989), BICUDO *ET AL.* (1992), SCHETTY (1998), SILVA (1999), PERES & SENNA (2000), FERRAGUT *ET AL. (*2005) and *TUCCI ET AL.* (2006).

LEITE (1974) is the floristic inventory of the Chlorococcales of the Parque Estadual das Fontes do Ipiranga located in the southern region of the Municipality of Sao Paulo. The author identified 60 taxa including species, varieties and taxonomic forms, 30 of which, or 50%, were representatives of the Scenedesmaceae. LEITE (1979) is, to date, the largest and best contribution to knowledge of the Scenedesmaceae family in the state of Sao Paulo. The work included all the material in LEITE (1974), extended the study area to the state of São Paulo and listed 118 infrageneric taxons of representatives of the order Chlorococcales, including 56 of Scenedesmaceae. SANT'ANNA (1984) is the formal publication of LEITE's work (1979), i.e. the doctoral thesis of the same author, then using her maiden name. This work contains 44 species representing the Scenedesmaceae family, namely: *Asterarcis quadricellularis* (Behre) Hegewald & Schmidle [as *Tetrastrum mitrae* (Tiwari & Pandey) Komarek], *Coronastrum anglicum* Flint, *Crucigenia fenestrata* (Schmidle) Schmidle, *Crucigenia quadrata* Morren, *Desmodesmus armatus* (Chodat) Hegewald var. *bicaudatus* (Guglielmetti) Hegewald [as *Scenedesmus bicaudatus* (Hansgirg) Chodat], *Desmodesmus communis (*Turpin) Hegewald [as *Scenedesmus quadricauda* (Turpin) Brebisson var. *maximus* West & West], *Desmodesmus denticulatus* (Lagerheim) An *et al.* (as *Scenedesmus denticulatus* Lagerheim var. *denticulatus* f. denticulatus*)*, *Desmodesmus opoliensis* (P. Richter) Hegewald (as *Scenedesmus opoliensis* P. Richter), Desmodesmus *perforatus (*Lemmermann) Hegewald (as *Scenedesmus perforatus* Lemmermann), *Desmodesmus protuberans* (Fritsch) Hegewald (as *Scenedesmus protuberans* Fritsch var. *protuberans* f. protuberans*)*, *Desmodesmus spinosus* (Chodat) Hegewald (as *Scenedesmus spinosus Chodat*), *Scenedesmus acuminatus* (Lagerheim) Chodat var. *acuminatus* f. acuminatus, *Scenedesmus acuminatus* (Lagerheim) Chodat var. *acuminatus* f *maximus* Uherkovich, *Scenedesmus acuminatus* (Lagerheim) Chodat var. *bernardii* (G.M. Smith) Dedusenko, *Scenedesmus acuminatus* (Lagerheim) Chodat var. *elongatus* G.M. Smith, *Scenedesmus acutus* Meyen var. acutus f. acutus, *Scenedesmus acutus* Meyen var. acutus f. *alternans* Hortobagyi, *Scenedesmus acutus* Meyen var. *acutus* f. *costulatus* (Chodat) Uherkovich, Scenedesmus *arcuatus* Lemmermann var. *arcuatus* f. arcuatus, *Scenedesmus arcuatus* Lemmermann var. arcuatus f. *arcuatus. gracilis* Hortobagyi, *Scenedesmus arcuatus* Lemmermann var. *arcuatus* f. *spinosus* Hortobagyi & Nemeth, Scenedesmus *bijugus* (Turpin) Kutzing var. *bijugus, Scenedesmus curvatus* Bohlin, Scenedesmus *decorus* Hortobagyi var. *bicaudato-granulatus* (Hortobagyi) Uherkovich, Scenedesmus *denticulatus* Lagerheim var. *australis* Playfair, *Scenedesmus denticulatus* Lagerheim var. *linearis* Hansgirg f. *costato-granulatus* (Hortobagyi) Uherkovich, Scenedesmus *denticulatus* Lagerheim var. *linearis* Hansgirg f. *granulatus* Hortobagyi, Scenedesmus *incrassatulus* Bohlin, Scenedesmus *naegelii* Brebisson, Scenedesmus *ovalternus* Chodat, *Scenedesmus protuberans* Fritsch var. *protuberans* f. *pologranulatus* Hortobabyi & Nemeth, *Scenedesmus quadricauda* (Turpin) Brebisson var. *parvus* G.M. Smith, *Scenedesmus quadricauda* (Turpin) Brebisson var. *quadrispina (*Chodat) G.M. Smith, *Scenedesmus quadricauda* (Turpin) Brebisson var. *westii* G.M. Smith,
Scenedesmus verrucosus Roll, *Scenedesmus wisconsinensis* (G.M. Smith) Chodat (as *Tetradesmus wisconsinensis* G.M. Smith), *Tetrallantos lagerheimii* Teiling, *Tetranephris brasiliense* Leite & C. Bicudo, *Tetrastrum elegans* Playfair, *Tetrastrum heteracanthum* (Nordstedt) Chodat, *Tetrastrum staurogeniaeforme* (Schröder) Lemmermann, *Westella botryoides* (W. West) De-Wildemann and *Willea irregularis (Wille)* Schmidle.

CARDOSO (1979) identified the phytophthora of the stabilization pond of the Cobertores Paraiba industry, in São José dos Campos, including 15 representatives of *Scenedesmus,* all with description and illustration, which are: *Desmodesmus armatus (*Chodat) Hegewald var. *bicaudatus* (Guglielmetti) Hegewald [as *Scenedesmus bicaudatus* (Hansgirg) Chodat], *Desmodesmus communis* (Turpin) Hegewald [as *Scenedesmus quadricauda* (Turpin) Brébisson var. *quadricauda],* *Desmodesmus denticulatus* (Lagerheim) An *et al.* (as *Scenedesmus denticulatus* Lagerheim var. *linearis* Hansgirg), *Desmodesmus opoliensis (*P. Richter) Hegewald (as Scenedesmus *opoliensis* P. Richter), *Desmodesmus spinosus* (Chodat) Hegewald (as *Scenedesmus spinosus* Chodat var. *spinosus],* *Didymogenes anomala* (G.M. Smith) Hindák [as *Scenedesmus anomalus* (G.M. Smith) Tiffany],

Pseudodidymocystis bicellularis (Chodat) Komárek (as Scenedesmus *bicellularis* Chodat), *Scenedesmus acutus* Meyen f. acutus, *Scenedesmus acutus* Meyen f. *alternas* Hortobágyi, *Scenedesmus acutus* Meyen f. *tetradesmiforme* (Woloszynska) Uherkovich, *Scenedesmus ecornis* (Ralfs) Chodat, Scenedesmus *incrassatulus* Bohlin, *Scenedesmus quadricauda* (Turpin) Brébisson var. *longispina* (Chodat) G.M. Smith f. *asymmetricus* (Hortobágyi) Uherkovich, *Scenedesmus quadricauda* (Turpin) Brébisson var. *quadrispina* (Chodat) G.M. Smith, *Scenedesmus spicatus* West & West and *Scenedesmus spinosus* Chodat var. *bicaudatus* Hortobágyi.

BiCUDO *ET AL*. (1992) described and illustrated *Crucigeniella crucifera* (Wolle) Komárek, *Desmodesmus protuberans* (Fritsch) Hegewald (as *Scenedesmus protuberans* Fritsch var. protuberans), *Desmodesmus communis* (Turpin) Hegewald [as *Scenedesmus quadricauda* (Turpin) Brébisson var. quadricauda], *Scenedesmus acuminatus* (Lagerheim) Chodat var. *acuminatus* f. *acuminatus* and *Scenedesmus ecornis* (Ralfs) Chodat var. *ecornis* in a study on the phytoplankton of the stretch of the Paranapanema River to be dammed for the construction of the Rosana Hydroelectric Power Station.

FERRAGUT *ET AL*. (2005) published the periphytic and planktonic phycoflora of the IAG Lake, located in the Parque Estadual das Fontes do Ipiranga, in the southern region of the Municipality of São Paulo, describing and illustrating the following 28 taxa classified among the Scenedesmaceae: *Crucigenia tetrapedia* (Kirchner) West & West, *Crucigeniella rectangularis* (Nägeli) Komárek, *Desmodesmus armatus* (Chodat) Hegewald var. armatus, *Desmodesmus armatus* (Chodat) Hegewald var. *bicaudatus* (Guglielmetti) Hegewald, *Desmodesmus communis* (Turpin) Hegewald [as *Scenedesmus quadricauda* (Turpin) Brébisson *"sensu"* Chodat], *Desmodesmus denticulatus* (Lagerheim) An *et al.* var. *linearis* (Hansgirg) Hegewald, *Desmodesmus intermedias* (Chodat) Hegewald var. *acutispinus* (Roll) Hegewald, Desmodesmus *intermedius* (Chodat) Hegewald var. *intermedius, Desmodesmus maximus* (West & West) Hegewald, Desmodesmus *opoliensis* (P. Richter) Hegewald var. *opoliensis, Desmodesmus polyspinosus* (Hortobágyi) Hegewald, Desmodesmus *pseudodenticulatus* (Hegewald) Hegewald, *Desmodesmus spinosus* (Chodat) Hegewald, *Pseudodidymocystis bicellularis* (Komárek) Hegewald & Deason [as *Didymocystis bicellularis* (Chodat) Komárek], *Pseudodidymocystis fina* (Komárek) Hegewald & Deason (as *Didymocystis fina* Komárek), *Scenedesmus acuminatus* (Lagerheim) Chodat var. *acuminatus, Scenedesmus acutus* Meyen var. *acutus* f. *acutus, Scenedesmus acutus* Meyen var. *acutus* f *alternans* Hortobágyi, *Scenedesmus danubialis* Hortobágyi, *Scenedesmus dimorphus* (Turpin) Kützing, Scenedesmus *ecornis* (Ehrenberg) Chodat, Scenedesmus *gutwinskii* Chodat, Scenedesmus *ellipticus* Corda [as *Scenedesmus linearis* Komárek], *Scenedesmus obliquus* (Turpin) Kützing, *Scenedesmus obtusus* Meyen var. *obtusus, Scenedesmus ovalternus* Chodat, *Tetrallantos lagerheimii* Teiling and *Tetrastrum triangulare* (Chodat) Komárek.

The above works by LEITE (1974, 1979), CARDOSO (1979), SANT'ANNA (1984), BICUDO *ET AL*. (1992) and FERRAGUT *ET AL*. (2005) include keys for identifying the genera, species, varieties and taxonomic forms identified, as well as comprehensive descriptions and plenty of illustrations.

The following 41 taxa of Scenedesmaceae are described and illustrated in SANT'ANNA *ET AL*. (1989): *Crucigenia fenestrata* (Schmidle) Schmidle, Crucigenia *quadrata* Morren, Crucigenia *tetrapedia* (Kirchner) West & West, *Crucigeniella crucifera* (Wolle) Komárek, *Crucigeniella rectangularis* (Nägeli) Komárek, *Desmodesmus armatus* (Chodat) Hegewald var. *bicaudatus* (Guglielmetti) Hegewald (as S. *bicaudatus* Dedusenko), *Desmodesmus communis* (Turpin) Hegewald [as *Scenedesmus quadricauda* (Turpin) Brébisson], Desmodesmus *denticulatus* (as *Scenedesmus denticulatus* Lagerheim), Desmodesmus *intermedius* (Chodat) Hegewald (as *Scenedesmus intermedius* Chodat), *Desmodesmus opoliensis* (P. Richter) Hegewald (as *Scenedesmus opoliensis* P. Richter), Desmodesmus *opoliensis* (P. Richter) Hegewald var. *carinatus* Lemmermann [as S. *opoliensis* P. Richter var. *carinatus* (Lemmermann) Chodat], *Desmodesmus opoliensis* (P. Richter) Hegewald var. *alatus* (Dedusenko) Hegewald (as S. *carinatus* Lemmermann var. *bicaudatus* Hortobágyi), *Desmodesmus protuberans* (Fritsch) Hegewald (as *Scenedesmus protuberans* Fritsch), *Desmodesmus serratus* (Bohlin) Hegewald [as .S' *serratus* (Corda) Bohlin], *Didymogenes anómala* (G.M. Smith) Hindák, *Didymogenes palatina* Schmidle, *Scenedesmus acuminatus* (Lagerheim) Chodat, *Scenedesmus acuminatus* (Lagerheim) Chodat f. *maximus* Uherkovich, *Scenedesmus acuminatus* (Lagerheim) Chodat var. *bernardii* (G.M. Smith) Dedusenko, *Scenedesmus acuminatus* (Lagerheim) Chodat var. *elongatus* G.M. Smith, *Scenedesmus acutus* Meyen, Scenedesmus. *acutus* Meyen f. *alternans* Hortobágyi, Scenedesmus *arcuatus* Lemmermann, *Scenedesmus arcuatus* Lemmermann f. *spinosus* Hortobágyi & Németh, *Scenedesmus bijugus* (Turpin) Kützing var. bijugus, Scenedesmus *bijugus* (Turpin) Kützing var. *disciformis* (Chodat) Leite, *Scenedesmus brevispina* (G.M. Smith) Chodat,

Scenedesmus *denticulatus* (Lagerheim) An *et al.* var. *australis* Playfair, Scenedesmus *ellipsoideus* Chodat, Scenedesmus *granulatus* West & West, *Scenedesmus ovalternus Chodat, Scenedesmus obtusus* Meyen [as Scenedesmus *ovalternus Chodat* var. *graeveniizii* (Bernard) Chodat], *Scenedesmus quadricauda* (Turpin) Brébisson var. *longispina* (Chodat) G.M. Smith f. *asymmetricus* (Hortobágyi) Uherkovich), *Scenedesmus verrucosus* Roll, *Tetrallantos lagerheimii* Teiling, *Tetrastrum elegans* Playfair, Tetrastrum *heteracanthum* (Nordstedt) Chodat, Tetrastrum *peterfii* Hortobágyi, *Tetrastrum triangulare* (Chodat) Komárek and *Westella botryoides* (W. West) De-Wildemann. These are phytoplankton from Lake Garbas and the study is mainly taxonomic, but also covers ecological aspects.

Susana Petersen Schetty has published a contribution to the planktonic phytoplankton of Lago das Ninféias, located in the Parque Estadual das Fontes do Ipiranga, in the southern region of the municipality of São Paulo. In this work, *Scenedesmus acutus* Meyen and *Desmodesmus protuberans* (Fritsch) Hegewald (as *Scenedesmus protuberans* Fritsch) are mentioned and illustrated (SCHETTY 1998).

In a study of the phytoplankton of Monte Alegre Lake, a eutrophic reservoir located in Ribeirao Preto, SILVA (1999) included the following 13 taxa of Scenedesmaceae: *Crucigenia tetrapedia* (Kirchner) West & West, *Crucigeniella pulchra* (West & West) Komárek, *Desmodesmus denticulatus* (Lagerheim) An *et al.* (as *Scenedesmus denticulatus* Lagerheim), *Desmodesmus opoliensis* (P. Richter) Hegewald (as *Scenedesmus opoliensis* P. Richter var. *danubialis* Hortobágyi), *Desmodesmus protuberans* (Fritsch) Hegewald (as *Scenedesmus protuberans* Fritsch & Rich), *Desmodesmus spinosus* (Chodat) Hegewald (as Scenedesmus *spinosus Chodat*), *Scenedesmus acuminatus* (Lagerheim) Chodat, *Scenedesmus arcuatus* (Lemmermann) Lemmermann var. *platydiscus* G.M. Smith, Scenedesmus *ellipticus* Corda, Scenedesmus *javanensis* Chodat, *Scenedesmus longispina* Chodat, *Tetrallantos lagerheimii* Teiling and *Tetrastrum heteracanthum* (Nordstedt) Chodat, with descriptions and illustrations of some of these species. The work aimed to identify the floristic composition of the phytoplankton community, analyze the complexity of taxa, compare it with that of other environments and use the taxa to identify the trophy of the lake.

PERES & SENNA (2000) illustrated the following 12 representatives of Scenedesmaceae: *Crucigenia fenestrata* Schmidle, *Desmodesmus armatus* (Chodat) Hegewald var. *bicaudatus* (Guglielmetti) Hegewald [as *Scenedesmus armatus* Chodat var. *bicaudatus* (Guglielmetti) Chodat], *Desmodesmus* cf. *denticulatus* (Lagerheim) An *et al.* (as *Scenedesmus* cf. *denticulatus* Lagerheim), *Desmodesmus hystrix* (Lagerheim) Hegewald (as *Scenedesmus hystrix* Lagerheim var. *hystrix),* *Desmodesmus protuberans* (Fritsch) Hegewald (as *Scenedesmus protuberans* Fritsch), *Desmodesmus communis* (Turpin) Hegewald [as *Scenedesmus quadricauda* (Turpin) Brébisson var *quadricauda],* *Dimorphococcus lunatus* A. Braun, *Scenedesmus armatus* (Chodat) Hegewald var. *boglariensis* Hortobágyi, Scenedesmus *bijugus* Chodat, *Scenedesmus* sp, *Tetrallantos lagerheimii* Teiling and *Tetrastrum* sp. in a study on integrated studies carried out in ecosystems of the Jataí Ecological Estate and, more specifically, the Diogo Lagoon.

Working with the phytoplankton of Lake Garbas, TUCCI *et al.* (2006) illustrated and described the following seven species of Scenedesmaceae: *Pseudodidymocystis fina* (Komárek) Hegewald & Deason, *Pseudodidymocystis planctonica* (Korsikov) Hegewald, Scenedesmus *indicus* Philipose, Scenedesmus *regularis* Swirenko, Scenedesmus *semipulcher* Hortobágyi, Scenedesmus *spinosus* Chodat and *Tetrastrum komarekii* Hindák.

The works of KLEEREKOPER (1937, 1939), BRANCO (1959, 1960, 1961, 1962, 1964), PALMER (1961), POTEL (1964), BRANCO ET AL. (1963), BICUDO & BICUDO (1967), XAVIER (1979), ROQUE (1980), TUNDISI & HINO (1981), XAVIER ET AL. (1985), SANT'ANNA *ET AL.* (1988), ROLLA ET AL. (1990), SCHWARZBOLD (1992), MARINHO (1994), BEYRUTH (1996), RAMÍREZ (1996), MOURA (1996), BICUDO ET AL. (1999), LOPES (1999), GENTIL (2000), SOUZA (2000), VERCELLINO (2001), CROSSETTI (2002), TUCCI (2002), BARCELOS (2003), CARVALHO (2003), BIESEMEYER (2005), FERREIRA (2005), FONSECA (2005), CERIONE *ETAL.* (2008) and RODRIGUES (2008) referred to the presence of Scenedesmaceae species in the state of São Paulo, but all of them had an ecological approach. In most of these works, the names of the materials identified are only listed, without description and/or illustration. Other works presented brief descriptions which often did not include the diagnostic characteristics of the respective taxa. Some of these papers have illustrations depicting the materials identified. BICUDO & BICUDO (1967) cited *Scenedesmus acuminatus* (Lagerheim) Chodat f. *maximus* Uherkovich, *Scenedesmus acuminatus* (Lagerheim) Chodat var. *tortuosus* (Skuja) Uherkovich and *Scenedesmus brasiliensis* Bohlin when studying the floating algae communities of Lago das Ninféias located in the Parque Estadual das Fontes do Ipiranga, in the southern region of the Municipality of

São Paulo. In this work, the authors discussed the ontogeny and floristic composition of these communities.

KLEEREKOPER (1937) illustrated *Scenedesmus bijuga* (Turpin) Lagerheim in a study on the biology of the old Santo Amaro reservoir (now the Guarapiranga reservoir). Two years later, KLEEREKOPER (1939) cited some species of Scenedesmaceae in a limnological study of the same reservoir.

BRANCO (1959) only mentioned the genus *Scenedesmus* in a paper on Sanitary Hydrobiology. The same genus was mentioned by the same author in another paper, this time on the hydrobiological problems resulting from the damming of the Cotia, Guarapiranga and Rio Grande reservoirs in the municipality of São Paulo (BRANCO 1961a). BRANCO (1961b) again mentioned the genus *Scenedesmus,* this time in a paper on the biology of the Alto Cotia reservoirs. And in 1962 and 1964, the same author again only mentioned the genus *Scenedesmus,* without illustrating or describing the material studied, making it impossible to re-identify it (BRANCO 1962, 1964). BRANCO *ET al.* (1963) published a manual for the identification of the main genera of algae that occur in the water supply of the state of Sao Paulo, gathering information on the importance of each one in the treatment of water and sewage. In this work, the genera *Crucigenia, Dimorphococcus and Scenedesmus are* represented by brief descriptions and illustrations.

PALMER (1961) and POTEL (1964) refer to the presence of the genus *Scenedesmus* in the water supply of the Municipality of Sao Paulo and its surroundings without, however, identifying the species, variety or taxonomic form or describing and/or illustrating the material examined. As happened before, given the lack of information in these works, it is impossible to re-identify this material.

XAVIER (1979) studied the seasonal variation of the phytoplankton of the Billings reservoir, in the municipality of São Paulo, and listed the species identified, but without describing them. There are, however, illustrations of *Desmodesmus opoliensis* (P. Richter) Hegewald (as *Scenedesmus opoliensis* P. Richter) and *Scenedesmus acuminatus* Lagerheim. In XAVIER *ET AL.* (1985), the authors made reference to some genera of the family, however, also without description and/or illustration of the material they examined.

TUNDISI & HINO (1981) published a list of species and their respective growth periods in the phytoplankton of the Lobo Reservoir (Broa), but without description or illustration, making it impossible to re-identify the material studied.

A qualitative study of the phytoplankton of the Serraría Reservoir was carried out by SANT'ANNA *ET AL.* (1988), where the authors described and illustrated 11 taxa representing Scenedesmaceae.

Various materials were identified down to genus level in the works on limnological and ecological aspects of the Scenedesmaceae family by ROQUE (1980) and ROLLA ET AL. (1990), respectively.

SCHWARZBOLD (1992) studied the effects of the flooding regime of the Moji Guacu river on the structure, diversity, production and stock of the periphyton of Lagoa do Inferno. The work mentioned and illustrated five taxa of representatives of the family.

MARINHO (1994) studied the dynamics of the phytoplankton community of a small shallow reservoir densely colonized by submerged aquatic macrophytes, Acude do Jacaré, located in Moji Guacu. The author listed and illustrated three species of *Scenedesmus: Desmodesmus brasiliensis* (Bohlin) Hegewald (as Scenedesmus *brasiliensis* Bohlin), *Scenedesmus ellipticus* Corda and *Scenedesmus javanensis* Chodat; and included descriptions of all three.

BICUDO *ET AL.* (1999) addressed the dynamics of phytoplankton populations in a eutrophicated environment, providing a very extensive list of species, but without including descriptions or illustrations to enable their re-identification. The species included in this work are: *Crucigenia fenestrata* (Schmidle) Schmidle, Crucigenia *quadrata* Morren, Crucigenia *tetrapedia* (Kirchner) West & West, Crucigeniella *crucifera* (Wolle) Komárek, *Crucigeniella rectangularis* (Nägeli) Komárek, *Desmodesmus abundans* (Kirchner) Chodat (as *Scenedesmus abundans* Kirchner), *Desmodesmus armatus* (Chodat) Hegewald var. *bicaudatus* (Guglielmetti) Hegewald (as *Scenedesmus bicaudatus* Dedusenko), *Desmodesmus communis* (Turpin) Hegewald [as *Scenedesmus quadricauda* (Turpin) Brébisson var. *longispina* (Chodat) G.M. Smith f. *asymmetricus* (Hortobágyi) Uherkovich], *Desmodesmus denticulatus* (Lagerheim) An *et al.* (as *Scenedesmus denticulatus* Lagerheim), *Desmodesmus opoliensis* (P. Richter) Hegewald var. *carinatus* Lemmermann [as *Scenedesmus carinatus* (Lemmermann) Chodat var. *carinatus* (Lemmermann) Chodat], *Desmodesmus hystrix* (Lagerheim) Hegewald (as *Scenedesmus hystrix* Lagerheim), *Desmodesmus intermedias* (Chodat)

Hegewald (as *Scenedesmus intermedias* Chodat), *Desmodesmus opoliensis* (P. Richter) Hegewald (as *Scenedesmus opoliensis* P. Richter), *Desmodesmus intermedius* (Chodat) Hegewald (as *Scenedesmus intermedius* Chodat), *Desmodesmus lunatus* (Chodat) Hegewald (as *Scenedesmus polyglobulus* Hotobágyi), Desmodesmus *opoliensis* (P. Richter) Hegewald (as *Scenedesmus opoliensis* P. Richter), *Desmodesmus protuberans* (Fritsch) Hegewald (as *Scenedesmus protuberans* Fritsch), *Desmodesmus serratus* (Bohlin) Hegewald [as *Scenedesmus serratus* (Corda) Bohlin], *Desmodesmus spinosus* (Chodat) Hegewald (as *Scenedesmus spinosus* Chodat), *Didymocystis inermis* (Fott) Fott, *Didymogenes anomalus* (G.M. Smith) Hindák [as *Scenedesmus anomalus* (G.M. Smith) Ahlstrom & Tiffany var. *acaudatus* Hortobágyi], *Didymogenes palatina* Schmidle, *Pseudodidymocystis planctonica* (Korsikov) Hegewald & Deason (as *Didymocystis planctonica* Korsikov), *Pseudotetrastrum punctatum* (Schmidle) Hindák [as *Tetrastrum punctatum* (Schmidle) Ahlstrom & Tiffany], *Scenedesmus acuminatus* (Lagerheim) Chodat var. *acuminatus* f *maximus* (Uherkovich) Ergasev, *Scenedesmus acuminatus* (Lagerheim) Chodat var. *elongatus* G.M. Smith, *Scenedesmus obliquas* (Turpin) Kützing var. *dimorphus* (Turpin) Kützing (as *Scenedesmus acutus* Meyen var. acutus), *Scenedesmus acutus* Meyen var. *acutus* f. *alternans* Hortobágyi, Scenedesmus *arcuatus* Lemmermann var. arcuatus, *Scenedesmus arcuatus* Lemmermann var. *arcuatus* f *spinosus* Hortobágyi & Németh, *Scenedesmus bijugus* (Turpin) Kützing var. bijugus, *Scenedesmus bijugus* (Turpin) Kützing var. *disciformis* (Chodat) Leite, Scenedesmus *brevispina* (G.M. Smith) Chodat, *Scenedesmus carinatus* (Lemmermann) Chodat var. *bicaudatus* Hortobágyi, *Scenedesmus denticulatus* Lagerheim var. *australis* Playfair, Scenedesmus *ellipsoideus* Chodat, Scenedesmus *granulatus* West & West, Scenedesmus *javanensis* Chodat [as *Scenedesmus acuminatus* (Lagerheim) Chodat var. *bernardii* (G.M. Smith) Dedusenko], *Scenedesmus obtusus* Meyen (as Scenedesmus *ovalternus Chodat*), *Scenedesmus ovalternus* Chodat var. *graeveniizii* (Bernard) Chodat, Scenedesmus *semipulcher* Hortobágyi, *Scenedesmus verrucosus* Roll, *Tetrallantos lagerheimii* Teiling, *Tetrastrum elegans* Playfair, Tetrastrum *heteracanthum* (Nordstedt) Chodat, Tetrastrum *peterfii* Hortobágyi, *Tetrastrum triangulare* (Chodat) Komárek and *Westella botryoides* (W. West) De-Wildemann.

GENTIL (2000) studied the seasonal variation of phytoplankton in a subtropical eutrophic lake, Lago das Garças, located in the Parque Estadual das Fontes do Ipiranga, in the southern region of the municipality of São Paulo, and some of its health aspects. In this work, the author described and illustrated 37 taxa of Scenedesmaceae.

Souza (2000) presented a list of 33 Scenedesmaceae taxa, but without description or illustration, making it impossible to confirm their identifications.

CARVALHO (2003) studied the phytoplankton community of reservoirs in the state of São Paulo with a view to biomonitoring these environments and made reference to three species of Scenedesmaceae, namely: *Scenedesmus quadricauda* Turpin [today *Desmodesmus communis* (Turpin) Hegewald], *Didymocystis planctonica* (Korsikov) Hegewald & Deason [today *Pseudodidymocystis planctónica* (Komárek) Hegewald & Deason] and *Crucigenia quadrata* Morren.

FERREIRA (2005) studied the community of periphytic algae adhered to *Eichhornia azurea* Kützing in two lagoons located on the banks of the Jurumirim Reservoir and provided a list of the 17 species of Scenedesmaceae without, however, describing or illustrating them. FERREIRA (2005) only cited the genera and species.

CERIONE *ET al.* (2008) carried out a floristic survey of planktonic algae species and provided information on the characteristics of the water in the Quinzinho de Barros Zoo Lake in Sorocaba. In this work, the authors simply listed three species of representatives of Scenedesmaceae, making it impossible to re-identify the species because they did not include a description and/or illustration: *Scenedesmus ecornis* (Ehrenberg) Chodat, *Desmodesmus communis* (Turpin) Hegewald [as *Scenedesmus quadricauda* (Turpin) Brébisson] and *Crucigenia quadrata* Morren.

RODRIGUES (2008) studied the diversity of cyanobacteria and algae in the Billings and Guarapiranga reservoirs, recording 10 taxa of Scenedesmaceae, as follows: *Crucigeniella crucifera* (Wolle) Komárek, *Desmodesmus armatus* (Chodat) Hegewald var. *bicaudatus* (Guglielmetti) Hegewald, *Desmodesmus communis* (Turpin) Hegewald, *Desmodesmus denticulatus* (Lagerheim) An *et al*, *Desmodesmus opoliensis* (P. Richter) Hegewald, *Scenedesmus acuminatus* (Lagerheim) Chodat, *Scenedesmus bernardii* G.M. Smith, *Scenedesmus disciformis* (Chodat) Fott & Komárek, *Tetrastrum homoiacanthum* (Huber-Pestallozi) Comas and *Westella botryoides* (W. West) De-Wildeman.

Several authors have carried out ecological studies in the Parque Estadual das Fontes do Ipiranga, working with the phytoplankton or periphyton of certain environments in the park. Thus,

CHAVES (1978), MOURA (1996), RAMÍREZ (1996), VERCELLINO (2001), CROSSETTI (2002), TUCCI (2002), BARCELOS (2003), FONSECA (2005) and FERMNO (2006) carried out their research in Lago das Garças, a eutrophic environment; LOPES (1999) and VERCELLINO (2001) in Lago do IAG, an oligotrophic environment; and BIESEMEYER (2005) and FONSECA (2005) in Lago das Ninféias, an oligo-mesotrophic environment. Of all these works, MOURA (1996) and TUCCI (2002) are the only ones to present illustrations of some species of Scenedesmaceae. The others only listed the materials identified.

Chapter 2
Objectives

The objectives of this research were as follows:

1. To inventory the taxonomic diversity of the Scenedesmaceae family in the area of the State of Sao Paulo.
2. Inventory the variability, at a population level, of the characteristics used in describing (diacritical, metric and meristic) species, varieties and taxonomic forms of Scenedesmaceae.
3. To evaluate the current use of these characteristics as diagnostics in the Scenedesmaceae family.
4. To provide subsidies for ecology, genetics, cytology, physiology, biochemistry, molecular biology, etc. projects that require prior knowledge of the taxonomic composition of the local phycological flora.

Chapter 3
Material and methods

3.1. Study area

The area covered by this state is the state of Sao Paulo, located in the southeast of Brazil (Fig. 3).

3.2. Material studied

In 1960, a program to collect algae material from continental and marine waters was started with the aim of producing the phycological flora of the State of São Paulo, which aimed to cover the entire area of the State. The result of this 40-year effort was the collection of around 3,500 samples of material from continental waters, which were incorporated into the collection of the State Scientific Herbarium "Maria Eneyda P. Kauffmann Fidalgo" (SP), of the Botany Institute of the Sao Paulo State Environment Secretariat. In the case of material from continental waters, the collection program gave priority to sampling in lentic environments as well as planktonic material. Consequently, several collections have been made recently, especially between 1999 and 2002, in order to cover the gaps in lotic environments and periphytic material mentioned above.

153 sample units collected in the state of Sao Paulo were examined. Of these, 55 were already part of the herbarium collection of the Botanical Institute and 98 were collected during the "Phycological Flora of the State of São Paulo" project under the BIOTA/FAPESP Program, Virtual Biodiversity Institute. Finally, 13 samples were collected more recently and, of these, six were included in the institutional collection after this study and seven were not because they did not contain microalgae. Representatives of Scenedesmaceae were detected in 91 localities. Each preparation was scrutinized to its maximum extent, i.e. by analyzing all the fields of all the possible horizontal transections of the preparation. The number of slides studied per sampling unit varied greatly, as they were prepared according to the richness observed in each sampling unit and as many preparations were examined as were necessary to taxonomically exhaust each sampling unit. Taxonomic exhaustion was accepted after 10 preparations had been made without any taxonomic novelties appearing.

Below, in ascending numerical order of their respective herbarium accession numbers, are the sampling units examined in which representatives of the Scenedesmaceae family were found. They are as follows.

The three maps of the state of São Paulo show, in the first, in red, the localities (municipalities) where specimens of Scenedesmaceae were found and, in blue, those where no representative of the family was found (Fig. 3). 3); in the second, in yellow, the distribution of representatives of this family according to the literature (Fig. 4); and in the third, in red, orange and yellow, the overlap of the occurrence of Scenedesmaceae in the state considering the information in the literature and in the samples examined (Fig. 5).

3.3. Material collection

The planktonic material was sampled using a net made from nylon fabric with a mesh size of 20 pm. The net was passed through the surface layer (± 30 cm) of the system as many times as necessary to obtain a reasonable amount of material. This reasonable quantity was measured with the naked eye by the presence of a greenish to greenish-brown mass accumulated by sedimentation at the bottom of the collection jar. This mucous-like mass is generally rich in algae.

In both lotic and lotic environments, sampling was carried out close to the shore, in the littoral zone, where there are usually floating and fixed aquatic plants that are totally or partially submerged.

These environments are natural concentrators of phytoplankton.

Periphytic material was sampled by manually squeezing submerged plants (bryophytes, pteridophytes and phanerogams) or their submerged parts and collecting the resulting liquid in a jar; or, when the size of the plant allowed, by collecting whole specimens or scraping off submerged parts of emerging individuals.

3.4. Fixing and preserving material

The materials were fixed and preserved immediately after collection, still in the field, with a 3-5% aqueous formalin solution. This procedure was due to the fact that the concentration of the material after collection very often accelerates the rate of cell division of certain species of algae and, consequently, the production of anomalous phenotypes. In fact, the daughter individuals (in this case, the newly formed cells) do not have time to develop all their structures in the same way as the mother cell before undergoing a new process of division. Since the apparent morphology of the cells forms the basis of the taxonomy of the Scenedesmaceae, it was imperative to ensure that such aspects resulting from malformations were avoided and, if they did occur, that they were not confused with expressions of intra-population morphological variation. Hence the need for immediate fixation of the material collected, in order to preserve the shape of the cells and their structures of taxonomic interest as close as possible to reality, i.e. the moment they were collected. In various circumstances, in addition to fixed material, live samples collected at PEFI were also observed, allowing the two types of material to be compared.

3.5. Inclusion of material in the institutional herbarium

All the material collected under the BIOTA/FAPESP Program, Virtual Biodiversity Institute was included in the State Scientific Herbarium "Maria Eneyda P. Kauffmann Fidalgo" (SP), of the Botanical Institute of the São Paulo State Department of the Environment.

The information on where these materials came from must follow the standard form for the BIOTA/FAPESP Program and the location of the collection sites was geo-referenced.

3.6. Examination of the material under the microscope

The materials were studied with the aid of a Zeiss Axioskop 2 optical microscope, between coverslip and coverslip, from the concentrated sample units.

The total length and maximum width of the cell, the angular spines and their length were measured directly using a digital micrometer eyepiece attached to the microscope's optical system.

3.7. Descriptive and illustrated material

The description of each identified taxon was provided as completely as possible, so as to include all diacritical, metric and meristic morphological characteristics of the vegetative stage. In addition, the variability of these characteristics at a population level was noted when possible. Below is a list of the characteristics studied:

1. Cell shape.
2. Size of the cenobium and cell identified by its maximum length (without the spines) and its maximum width (without the spines).
3. Ratio between the maximum length (without the spines) and the maximum width (without the spines) of the cell.
4. Length of spines.
5. Cell wall decoration.

The measurements are represented by their maximum and minimum metric limits. Extreme measurements, which fell outside the limits of gradual variation, are shown in parentheses. We interpreted the size of the spines not in absolute terms, but in relation to the maximum width of the cell, as follows:

$$(1) < 1/10 \text{ of the cell width} = \text{short spines.}$$
$$(2) > 1/3 \text{ and} < 1/2 \text{ of the cell width} = \text{medium spines.}$$
$$(3) > 1/2 \text{ the width of the cell} = \text{long spines.}$$

It should be borne in mind, however, that these limits have not always been easy to recognize in practice, given the extreme variation of these structures in population samples.

The descriptions of the taxa whose material examined appears as "only material from the literature" were based on the authors' own descriptions of the literature from the state of São Paulo, but where possible adding characteristics that could be interpreted from the illustrations for standardization purposes.

The illustrations were made with the intention of demonstrating the variability observed in each sampling unit and between the different sampling units examined. All the material illustrated is represented by its frontal view, which is essential for the taxonomic identification of the materials. The original drawings were made using a camera coupled to the microscope's optical system, in the form of pencil sketches. The most representative drawings were then selected and distributed on boards according to the order of citation of the taxa in this work. Finally, the drawings on the boards were finished in ink.

3.8. Taxonomic identification of the material

The identifications were based on the analysis of as many individuals as possible and, whenever permitted, on the analysis of populations due to the great morphological and metric variability presented by the individuals. This procedure was designed to increase the validity of the interpretations. The identifications were provided on the basis of specialized bibliography and preferably used works of the flora, monograph and review types, both classic and recent. Works of lesser taxonomic scope relating to one or other species were also used in the identification process. In the latter case, the most recent works available were consulted.

3.9. Classification system used

The family Scenedesmaceae is classified according to VAN-DEN-HOEK *ETAL.* (1997) and KOMÁREK & FOTT (1983) in the Division Chlorophyta, Class Chlorophyceae and Order Chlorococcales. This was the classification system adopted in this study.

3.10. Taxonomic identification key

An artificial dichotomous key was prepared to identify the species, varieties and taxonomic forms currently inventoried. We opted for a key of paired excluding alternatives to cater for beginners as well. These alternatives were based on the morphological, metric and meristic characteristics of the materials from the state of Sao Paulo.

3.11. Bibliographical references

The citations of the bibliographical references at the end of the work are in accordance with the rules for publication in the journal "Hoehnea" of the Botanical Institute. The titles of scientific journals and books are given in full, also in accordance with the requirements of "Hoehnea" for the preparation of papers to be submitted for publication.

3.12. Literature material

All the material in the specialized literature from the state of Sao Paulo was evaluated. The taxonomic identifications of all the materials were reviewed, especially those that were described including their diagnostic characteristics, measured and/or illustrated. The identifications of materials deposited in herbaria, even if unofficial, were also re-evaluated. Whenever possible, materials included in the document-collections of aquatic ecology laboratories were examined. The other materials of which there was not enough information to re-evaluate them, or which were not found in herbaria and document collections and/or which were in exsiccates but could not be handled, were included in this work in the chapter entitled "Excluded taxa".

All bibliographical references that appear in theses, dissertations and/or course completion monographs and that only list species, without description or illustration, have been cited in the introduction to this study.

Chapter 4
Results and discussion

Subfamily Coronastroideae *sensu* Komárek & Fott 1983

Cells largely ovoid, ellipsoidal or lunate, united in more or less square cenobia, whose longitudinal axes of the cells are perpendicular or inclined in relation to the plane of the cenobium; cells united by delicate gelatinous appendages located at one or both cell poles or scattered throughout the cell wall; parietal chloroplastid, with or without pyrenoid; reproduction by 4 autopores united in 1 autocenobium which can form syncenobia.

Coronastrum Thompson 1938

Colonial individuals that float freely in the water. Cenobia formed by 4 cells distributed in a flat square. The cells can be globose, ellipsoid, ovoid, pyriform or lunate and are held together by remnants of the mother cell wall which appear as short strands of mucilage. In addition, each cell has a fragment of the mother cell wall in the shape of a tiny cap on the side facing the outside of the cenobium. The longitudinal axes of the cells are parallel to each other and the aforementioned caps or fins are arranged along these axes. Chloroplastidium parietal, laminar, with 1 pyrenoid (Bicudo & Menezes 2006).

Only one species identified:

***Coronastrum anglicum* Flint** (Fig. 6) Hydrobiologia2: 235, fig. 4A-D. 1950.

Cenobium quadratic, flat, formed by 4 cells, 1 thorn in each cell, sometimes forming multiple cenobium; cells oval, connected to each other by mucilaginous processes situated at more or less right angles to the major axis of the cells; cell length 4.6-5.7 pm, width 4.0-4.8 pm; chloroplastidiopoculiform, 1 pyrenoid.

Geographical distribution in the State of Sao Paulo

IN LITERATURE: **Municipality of Santo André** (SANT'ANNA 1984: 174, fig. 109110).

EXAMINED MATERIAL: **Municipality of Sao José do Barreiro**, SP-64, km 0.8, Queluz-Sao José do Barreiro road, marsh, with *Thypha* and Cyperaceae, 21-XI-1989, *A.A.J. de Castro & C.E.M.Bicudo* (SP188322).

Comment

The cells of *Coronastrum anglicum* Flint are connected to each other by mucilaginous processes located more or less at right angles to the major axis of the cells. *Coronastrum anglicum* Flint is morphologically very similar to *Coronastrum ellipsoideum* Fott, however, the latter species always has four-celled cenobia and never forms multiple cenobia (FOTT 1946). In addition, the latter species has cenobia formed by ellipsoid cells. However, SANT'ANNA (1984) considered the specimens she examined to be very similar to those of *C. ellipsoideum* Fott. In fact, she found both types of cenobium in the samples analyzed, identifying both, however, as representatives of *C. anglicum* Flint.

In the samples analyzed here, only individuals with four-celled cenobiums were found, although SANT'ANNA (1984) mentions that he found another type of cenobium, in this case multiple.

Coronastrum anglicum Flint was found in a single locality in the municipality of Sao José do Barreiro. Individuals were found that were very characteristic of the species and very similar to those illustrated in KOMÁREK & FOTT (1983).

Subfamily Crucigenioideae *sensu* Komárek & Fott 1983

Flat cenobiums with 4 cells arranged in a cross forming 2 pairs, one above the other in the same plane, the 2 inner cells situated more or less parallel to the 2 outer ones, joined obliquely to them by their ends; smooth cell wall, with warts or thorns; parietal chloroplastidium with or without pyrenoid; reproduction by autospores united in autocenobiums

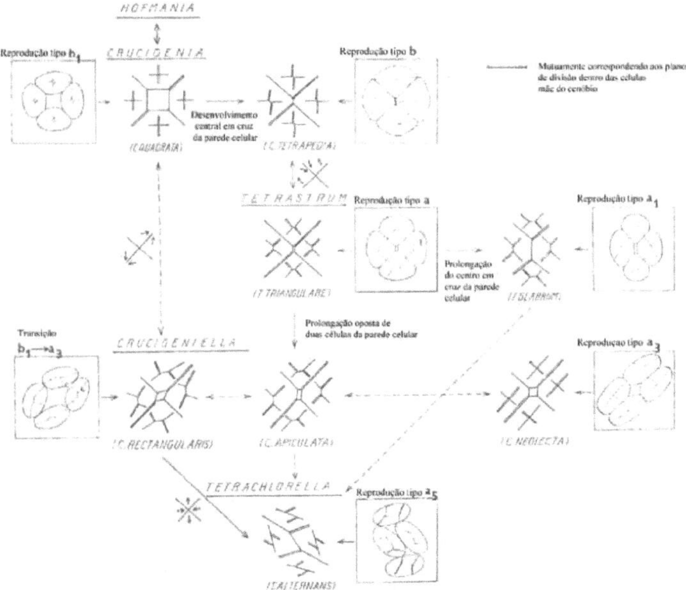

Figure 1. Cell division plans of some genera of the subfamily Crucigenoideae: *Crucigenia, Crucigeniella, Tetrastrum* and *Tetrachlorella* (KOMÁREK 1974).

Key for identifying the genera studied:

1. Cenobiums composed of pairs of cells located in the same plane, elongated in 1 direction. whose longitudinal axis passes through the pairs of cells2
2. Cenobiums with 2 cells more or less parallel or slightly converging with each other. the other 2 cells obliquely joined to the ends of the previous pair. or 4 cells arranged crosswise in more or less square cenobiums3
   3. Cenobium slightly surrounded by remains of the maternal cell wall; reproduction usually by 2-4 autospores oriented perpendicularly on the longitudinal axis of the cenobium; they form permanent syncenobia... *Willea*
   4. Cenobiums not surrounded by remains of the maternal walls; usually produced by 4 autospores oriented in the same direction as the maternal cenobium; sometimes form transient syncenobia..*Crucigeniella*
5. Cenobia with 2 more or less parallel, slightly converging cells to each other and the other 2 cells obliquely joined at the ends of the previous pair... *Tetrachlorella*
6. More or less cross-shaped scenes, square or extended in 1 .... direction4
   7. More or less square cenobiums, with 1 small space in the center; cell wall not ornamented; daughter cenobia forming a 45° angle to the mother cenobium..*Crucigenia*
   8. Cenobiums generally elongated in 1 direction, rarely square; smooth cell wall or with granules, warts or spines; daughter cenobiums oriented in the same direction as the mother cenobium5
9. Smooth cell wall; daughter cenobia often joined by remnants of maternal cell walls *Westella*
10. Smooth cell wall, with spines on the outer margins of the cells or with granules or warts on the surface6
    11. Cenobium formed by cells arranged in 2 pairs, more or less in level..........*Didymogenes*
    12. Transient syncenobes (if formed) that are not joined by remains of the maternal walls *Tetrastrum*

Crucigenia Morren 1830

Flat cenobiums made up of four cells that float freely in the environment. The shape of the cenobium varies from almost circular to square, rectangular and rhombic. The cells can be ellipsoidal, triangular, trapezoidal or quarter-circular and are distributed around a small central space. Multiple cenobia are common. Lateral chloroplastidium, blade-shaped, with or without pyrenoid (BICUDO & MENEZES 2006).

Key to identifying the species studied: 1. Cenobia with ..... convex outer cell margins2
1. Cenobia with more or less straight or slightly concave outer cell margins3
   2. Smooth cell wall; pyrenoid present...*C. quadrata*
   3. Ornate cell wall with wart-like thickenings;
      pyrenoid present ....*C. mucronata*
4. Opening in the center of the cenobium present...*C. fenestrata*
5. Aperture in the center of the cenobium absent..*C. tetrapedia*

Crucigenia fenestrata (Schmidle) Schmidle (Fig. 7)

Allgemeine Botanische Zeitschrift 6: 234. 1901.

Basionimo: *Staurogenia fenestrata* Schmidle, Allgemeine Botanische Zeitschrift 3: 107. 1897.

Cenobium flat, formed by 4 cells, cells arranged in a cross, with 1 small spire in the center of the cenobium; trapezoidal cells, with slightly concave outer margins; cell length 4.0-10.8 pm, width 4.2-7.5 pm; parietal chloroplastid, no pyrenoid.

Geographical distribution in the State of Sao Paulo

IN LITERATURE: **Municipality of Luiz Antonio** (PERES & SENNA 2000: 473, fig. 5). **Municipality of Avai, Pindamonhangaba** and **Sao Paulo** (SANT'ANNA 1984: 177, fig. 112). **Municipality of Juquiá** [SANT'ANNA 1984: 263, fig. 178, as *Tetrastrum mitrae* (Tiwari & Pandey) Komárek]. **Municipio de Juquiá** (SANT'ANNAETAL. 1988: 91, fig. 36). **Municipality of Sao Paulo** (SANT'ANNAETAL. 1989: 95, fig. 59).

MATERIAL EXAMINED: only material from the literature.

Comment

According to SANT'ANNA (1984), *Crucigenia fenestrata* (Schmidle) Schmidle is characterized by the trapezoidal shape of the cells, the quadratic spire in the center of the cenobium and the absence of a pyrenoid. These features were observed in the specimens studied by SANT'ANNA ET AL. (1989). According to HINDÁK (1984a), *Crucigenia fenestrata* (Schmidle) Schmidle can be distinguished from *C. tetrapedia* (Kirchner) West & West by having a square intercellular space with concave sides, whereas *C. tetrapedia* (Kirchner) West & West has a small rectangular intercellular space or this space may be missing. According to MARTINS-DA-SILVA (1996), *C. fenestrata* (Schmidle) Schmidle differs from other species of the genus in that the outer side of the cells is straight or slightly concave and it has no pyrenoid. We agree with the latter author, stating that the specimens from the state of Sao Paulo currently studied have characteristics identical to those she mentions.

KOMÁREK (1974) mentioned that, given the great variability of the central opening of the cenobium, it is not yet sufficiently clear whether *C. fenestrata* (Schmidle) Schmidle is identical and therefore synonymous with *C. tetrapedia* (Kirchner) West & West or whether it fits in with other species.

In SANT'ANNA (1984) it was identified as *Tetrastrum mitrae* (Tiwari & Pandey) Komárek. However, on re-examining the illustration (SANT'ANNA 1984: 263, fig. 178) we concluded that it was *C. fenestrata* (Schmidle) Schmidle, since the specimen shows all the diagnostic features of this species.

Crucigenia mucronata (G.M. Smith) Komárek (Fig. 8)

Archiv fur Protistenkunde 116: 25, fig. 31. 1974.

Cenobium flat, formed by 4 irregular cells, quadrangular cenobium with 1 quadrangular space in the center; oval to trapezoid cells, cell wall with 1 wart-like thickening at the angles where the cells contact, outer side of the cells slightly concave; cell length 4.5-8.28 pm, width 2.05-6.45 pm; parietal chloroplastidium, 1 pyrenoid.

Geographical distribution in the State of Sao Paulo

IN LITERATURE: **Municipio de Ribeirao Preto** [SILVA 1999: 286, fig. 33, as *Crucigeniellapulchra* (West & West) Komárek].

EXAMINED MATERIAL: **Municipality of Barretos**, Barretes, in the city, lake region, with grasses and Cyperaceae, 28-II-1990, *L.H.Z. Branco* (SP255772). **Municipality of Guaratinguetá**, Clube dos 500, lake, 01-IV-1966, *C.E.M. Bicudo* (SP96965). **Municipality of Itu**, SP-280, km 77, lake, 11-V-1977, *C.R. Leite* (SP139733). **Municipality of Pindamonhangaba,** BR-116, km 270-271, pond, 21-V-1966, *C.E.M. Bicudo* (SP96949); Sao Joao farm, Sao Joao pond, 21-V-1966, *C.E.M. Bicudo* (SP96950). **Municipality of Sao Carlos,** SP-310, km 222, Aldeia Conde do Pinhal, stream, 10-V-1973, *C.E.M. Bicudo & L. Sormus* (SP104723). **Municipio de Sao Paulo,** Parque Estadual das Fontes do Ipiranga, Lago do IAG, 03-VIII-1998, *I.S. Vercellino,* 23°38'08"S and 23°40'18"S, 46°36'48"W and 46°38'00"W (SP399779); Lago das Garbas, 12-VII-2006, *M. Borduqui,* 23°38'08"S and 23°40'18"S, 46°36'48"W and 46°38'00"W (SP399781); 17-I-2007,*M Borduqui, 23°38'08"S* and 23°40'18"S, 46°36'48"W and 46°38'00"W (SP399782). **Municipio de Sao Pedro do Turvo,** BR-153, 10 km from the municipality border, marsh, metafiton, 28-III-2001, *C.E.M. Bicudo, L.A. Carneiro & S.M.M. Faustino,* 22°48'46,3"S, 49°47'24,8"W, conductivity 60 pS cm⁻¹, pH 6,2 (SP355390).

Comment

Crucigenia mucronata (G.M. Smith) Komárek is similar in morphology to *Crucigeniella pulchra* (West & West) Komárek, but differs in that the cenobia are quadratic in shape in *Crucigenia mucronata* (G.M. Smith) Komárek and more elongated, with the cells projecting in relation to the longitudinal axis, in *Crucigeniella pulchra* (West & West) Komárek. According to KOMÁREK (1974), *Crucigenia mucronata* (G.M. Smith) Komárek can form syncenobia with up to 16 cells and *Crucigenia fenestrata* (Schmidle) Schmidle var. *tetraverruca* Hortobágyi is most likely a taxonomic variety of *Crucigenia mucronata* (G.M. Smith) Komárek. ROSA & OLIVEIRA (1990) mentioned that the specimens they examined from the state of Rio Grande do Sul showed lateral and pyrenoid protuberances, as well as the common presence of remains of the mother cell wall and cenobia made up of up to 32 cells. KOMÁREK (1983) observed that *Crucigenia mucronata* (G.M. Smith) Komárek generally occurred together with *Crucigeniella pulchra* (West & West) Komárek. This author stated that the taxonomic relationship between the two species is not very clear and that it is possible that there are morphological expressions (ecomorphs) in *Crucigeniellapulchra* (West & West) Komárek.

The samples currently analyzed showed that *Crucigenia mucronata* (G.M. Smith) Komárek is easily found in the state of Sao Paulo. The species was collected from six different localities. Some very characteristic individuals of the species were seen, but some specimens were confused with *Crucigeniella pulchra* (West & West) Komárek and *Crucigeniella crucifera* (Wolle) Komárek, despite the fact that this second species is not morphologically so close to *Crucigenia mucronata* (G.M. Smith) Komárek as *Crucigeniella pulchra* (West & West) Komárek, which also has individuals very similar to those of *Crucigenia mucronata* (G.M. Smith) Komárek.

***Crucigenia quadrata* Morren** (Fig. 9)

Annales des SciencesNaturelles 20: 415, pl. 15, fig. 1-5. 1830.

Cenobium flat, formed by 4 cells, cenobium irregularly quadrangular, with 1 quadrangular space in the center; cells elliptical to trapezoidal, outer margins markedly convex; cell length 6.0-9.28 pm, width 3.25-3.95 pm; parietal chloroplastid, 1 pyrenoid.

Geographical distribution in the State of Sao Paulo

IN LITERATURE: **Municipalities of Guaratinguetá, Itirapina, Moji das Cruzes, Rio Claro, Santo André, Sao Carlos, Sao Paulo, Sorocaba and Tambaú** (SANT'ANNA 1984: 178, fig. 111). **Municipality of Juquiá** (SANT'ANNA *ET AL.* 1988: 91, fig. 37). **Municipality of Sao Paulo** (SANT'ANNA ET AL. 1989: 95, fig. 63).

EXAMINED MATERIAL: **Municipality of Tambaú,** Clube de Tambaú, dam, 23- VI-1973, *D.M. Vital* (SP113574).

Comment

Crucigenia quadrata Morren is characterized by the elliptical to trapezoidal shape of the cell and the convex outer wall, in addition to the lateral walls being flattened by mutual compression (SANT'ANNA 1984).

AHLSTROM & TIFFANY (1934) considered *C. quadrata* Morren var. *octogona* Schmidle to be synonymous with *Tetrastrum alpinum* Schmidle. *Crucigenia quadrata* Morren var. *octogona* Schmidle f. *pulchra* Hortobágyi has a cenobium made up of four cells very similar to those of the standard variety of *C. quadrata* Morren. According to KOMÁREK (1974), the nomenclatural type in HORTOBÁGYI (1967) has a reproductive mode identical to that of representatives of *Tetrastrum,* making it necessary to consider *Crucigenia quadrata* Morren var. *octogona* Schmidle f. *pulchra* Hortobágyi a variety of *Tetrastrum triangulare* (Chodat) Komárek or an independent species.

Few individuals (only four) of this species have been collected in the state of Sao Paulo and only in one municipality. The individuals examined here are very typical of the species and no difficulty was encountered in identifying them.

Crucigenia tetrapedia **(Kirchner) West & West** (Fig. 10)
Transactions of the Royal Irish Academy: ser. B, 32: 62. 1902.

Cenobium flat, more or less square, without central spire, formed by 4 cells; triangular cells, with straight or slightly concave outer margins; cell length 2.2-5.0 pm, width 1.9-4.3 pm; parietal chloroplastidium, without pyrenoid.

Geographical distribution in the State of Sao Paulo

IN LITERATURE: **Municipality of Sao Paulo** (SANT'ANNA ETAL. 1989: 95, fig. 58; GENTIL 2000: 52, fig. 4; *FerragutETAL.* 2005: 152, fig. 73). **Municipality of Ribeirao Preto** (SILVA 1999: 286, fig. 30).

EXAMINED MATERIAL: **Municipality of Orlándia**, highway Morro Agudo towards Sao loaquim da Barra, km 13, dam, 29-V-2000, *C.E.M. Bicudo & D.C. Bicudo* (SP355380). **Municipality of Pindamonhangaba,** BR-116, km 270-271, lagoon, 21-V-1966, *C.E.M. Bicudo* (SP96949); Sao loao farm, Sao loao lagoon, 21-V-1966, *C.E.M. Bicudo* (SP96950). **Municipality of Pirassununga**, SP-225, km 23, Cascalbo neighborhood, lake, ?-XII-1973, *P.A.C. Sema* (SP123900). **Municipio de Sao Paulo**, Parque Estadual das Fontes do Ipiranga, Lago do IAG, 03-VIII-1998, *I.S. Vercellino,* 23°38'08"S and 23°40'18"S, 46°36'48"W and 46°38'00"W (SP399779).

Comment

More or less square cenobiums made up of triangular cells, without a spire in the center, a characteristic that will separate *Crucigenia* from *Tetrastrum*. Another typical feature of this species is the parietal chloroplastid without pyrenoid, in addition to being a cosmopolitan species.

According to SANT'ANNA *ET AL.* (1989) and SILVA (1999), the specimens studied show the same characteristics as the original illustration in WEST & West (1902), however, the illustration by FERRAGUT *ET AL.* (2005) shows a very small gap between the cells, which shows that this characteristic varies within the species. In fact, according to KOMÁREK (1974), there is no opening in the center of the cenobium, but only a small gap in which the cells touch each other more or less at the junction of the cell walls of the cenobium cells.

WEST & WEST (1902) made a valid nomenclatural combination *(Crucigenia tetrapedia)* even though they illustrated *C. fenestrata,* obviously another organism.

The species was common in the state of Sao Paulo, since it was collected in five municipalities. We also had no difficulty identifying it because the cenobiums have triangular cells. The only thing to watch out for is the opening in the center of the cenobium, which could confuse its identification with *Tetrastrum triangulare* (Chodat) Komárek, as the species is very similar in terms of the morphology of the cenobium and its cells.

Table 1. Differential characteristics between *C. fenestrata, C. mucronata, C. quadrata* and *C. tetrapedia.*

Species	External cell margin	Pyrenoid	Cell wall	Opening in the center of the cenobium
C. quadrata	convex	present	Lisa	square sword in the center of the cenobium
C. mucronata	slightly concave	present	with wart-like thickening	square sword in the center of the cenobium

C. fenestrata (Sant'Anna 1984)	more or less straight or slightly concave	absent	Lisa	small square sword in the center of the cenobium
C. tetrapedia	more or less straight or slightly concave	absent	Lisa	absent

Crucigeniella Lemmermann 1890

Cenobia formed by four flat, free-living, square to slightly rectangular cells, which can join together to form multiple cenobia. The cells can be ellipsoidal, reniform or even slightly asymmetrical in both the equatorial and lateral planes and the poles are broadly rounded or asymmetrically acuminate. In the central portion of the colony there is a rhomboidal spire. Lateral chloroplastid with a pyrenoid (BICUDO & MENEZES 2006).

Key to identifying the species studied:

1. Cenobium with .... convex outer cell margins2
1. Cenobium with almost straight outer cell margins
   or slightly concave .... *C. crucifera*
   2. Cells with pyrenoid .... *C. apiculata*
   2. .... Cells without pyrenoid
(sometimes not very visible) .... *C. rectangularis*

Crucigeniella apiculata (**Lemmermann**) **Komárek** (Fig. 11)
Archiv fur Protistenkunde 116: 38, fig. 67-69. 1974.

Flat cenobiums formed by 4-16 cells, square spire in the center, syncenobiums can reach 28 or more cells; asymmetrical oval cells, joining zone between cells occupying half the cell length, contact of cells asymmetrically acute, truncated, outer margins of cells more or less convex, distal cell ends forming 1 acute angle; cell length 4.2-5.8 pm, width 2.5-2.9 pm; parietal chloroplastidium, 1 pyrenoid.

Geographical distribution in the State of Sao Paulo

IN THE LITERATURE: nothing. First mention of the occurrence of the species in the state of Sao Paulo.

EXAMINED MATERIAL: **Municipality of Santa Albertina**, vicinal vereador ítalo Biani, 15 km after the city, stream, benthos, 24-IV-2001, *C.E.M. Bicudo, D.L. Costa & S.M.M. Faustino,* 20°3'20,1"S, 50°46'0,1"W, conductivity 110 pS cm^{-1} , pH 7,7 (SP355385). **Municipio de Sao Paulo**, Parque Estadual das Fontes do Ipiranga, Lago das Garras, 17-I-2007, *M. Borduqui,* 23°38'08"S, 23°40'18"S, 46°36'48'W and 46°38'00'W (SP399782).

Comment

The cells may show a thickening of the cell wall (KOMÁREK & FOTT 1983). According to COMAS (1996), this apical thickening of the cell wall is not as evident in Cuban specimens as it is in those from Central Europe. We agree with this author, as the current specimens from the state of Sao Paulo also show no thickening of the cell wall. Cenobia can reach 64 cells or more, however, they have shown syncenobia with a maximum of 28 cells. *Crucigeniella apiculata* (Lemmermann) Komárek was only found in two municipalities in the state of Sao Paulo, i.e. Santa Albertina and Sao Paulo. In addition, not many individuals of the species have been found and it is therefore currently considered a rare occurrence. This is the first mention of the species in the state of Sao Paulo. Finally, the individuals found were considered to be very typical of the species and there were no difficulties in identifying them.

Crucigeniella crucifera (**Wolle**) **Komárek** (Fig. 12)
Archiv fur Protistenkunde 116: 39, fig. 78-81. 1974.

Flat cenobia formed 4-16 cells, cenobia approximately quadratic, central space rhomboidal to quadratic; cells irregularly oblong, outer margin of cell almost straight or slightly concave; length of cell 3.1-9.2 pm, width 2.3-4.5 pm; parietal chloroplastid, 1 pyrenoid.

Geographical distribution in the State of Sao Paulo

IN LITERATURE: **Municipality of Sao Paulo** (SANT'ANNA *ET AL.* 1989: 95, fig. 6164; GENTIL 2000: 53, fig. 5). **Municipality of Teodoro Sampaio** (BICUDOETAL. 1992: 298, fig. 30-31).

EXAMINED MATERIAL: **Município de Cosmorama**, SP-320, km 496, acude, metafiton 24-IV-2001, *C.E.M. Bicudo, D.L. Costa & S.M.M. Faustino,* 20° 30'18,4"S, 49°46'14,4"W, conductivity 30 pS cm⁻¹ , pH 6,4 (SP355389). **Municipality of Itu**, SP-280, km 77, lake, 11-V-1977, *C.R. Leite* (SP139733).

Comment

According to SANT'ANNA ETAL. (1989), the species has approximately hexagonal cenobiums, with a rhomboidal to quadratic spire in the center. The cells are reniform and generally have a thickening at the outer edge. BICUDO etAL. (1992) observed only young specimens, which differ from adults in terms of the more reniform shape of the cells and the plane of division of the autospores in relation to the maternal cenobium. These authors mentioned that these morphotypes coincided with certain stages in the development of *Crucigeniella crucifera* (Wolle) Komárek, as illustrated in KOMÁREK (1974: figs. 78, 81). We agree with BicudoetAL. (1992), although the present specimens show cells with more oblong shapes, somewhat irregular, but still oblong.

Crucigeniella crucifera (Wolle) Komárek has currently only been found in two municipalities in the state of São Paulo. However, large populations and individuals very typical of the species have been recorded. No difficulties were encountered in identifying the species.

Crucigeniella rectangularis (Nageli) Komárek (Fig. 13)

Archiv fur Protistenkunde 116: 37, fig. 65-66. 1974.

Cenobium flat, formed of 4-16 cells, asymmetrical oval cells, with 1 irregular square opening in the center of the cenobium; outer cells slightly convex, with their cells delicately joined by a common mucilage; cell wall smooth, without thickening; cell length 4.0-8.0 pm, width 2.5-5.0 pm; pyrenoid sometimes not very visible.

Geographical distribution in the State of Sao Paulo

IN LITERATURE: **Municipality of Juquiá** (SANT'ANNA *ET AL.* 1988: 91, fig. 38). **Municipality of Sao Paulo** (SANT'ANNA ETAL. 1989: 95, fig. 62; FERRAGUTETAL. 2005: 152, fig. 74).

MATERIAL EXAMINED: only material from the literature.

Comment

According to KOMÁREK (1974), the exsiccates of *Crucigenia rectangularis* (Nageli) Komárek gave the specimens collected in Switzerland different diagnoses from the original and consequently enabled different interpretations (KOMÁREK 1974: fig. 65). The material illustrated by KOMÁREK (1974: fig. 102) has a cenobium with a rhomboid central opening, no pyrenoid and irregularly situated cells (compare with *Willea).* According to KOMÁREK (1983), *C. rectangularis (*Nageli) Komárek has been described in many ways. The diacritical characteristics of this species are the following: (7) ovoid or oval cells, (2) a distal polar thickening in each cell, *(3)* the shape of the central opening in the cenobium and *(4)* irregular number of autospores (cenobiums consisting of two cells also occur, as in *Willea)* (KOMÁREK 1983). PICELLI-VICENTIM (1987) mentioned that *C. rectangularis (*Nageli) Komárek resembles *Crucigenia quadrata* Morren, but the latter species always has square cenobiums, while *C. rectangularis* (Nageli) Komárek has rectangular ones, i.e. clearly elongated in one direction.

Crucigeniella rectangularis (Nageli) Komárek has ovoid, asymmetrical cells with an irregular square central space and can form syncenobia with 32-64 cells.

Table 2. Distinguishing features between C. *apiculata,* C. *crucifera* and C. *rectangularis.*

Species	Pyrenoid	External cell margin	Central opening of the cenobium	Cell wall
C. apiculata	present	convex	square	With or without thickening

C. crucifera	present	almost straight or slightly concave	rhomboidal to quadratic	no thickening
C. rectangularis (Ferragut *et al.* 2005)	absent	convex	irregular quadratic	smooth, without thickening

Didymogenes Schmidle 1905

Flat cenobiums, devoid of mucilage, made up of 2-4-8-16 cells, with solitary cells being uncommon. Cenobium cells do not grow tightly together, but are only close to each other, one higher up and the next lower down, in a parallel arrangement. The cells are semi-lunate, the cell wall can be smooth or warty, with or without terminal spines. Reproduction is by autospores (CARDOSO 1979).

Key to identify the species studied:

1. Outer cells with spines.. *D. anomala*
1. Outer cells without spines... *D. palatina*

Didymogenes anómala (G.M. Smith) Hindák (Fig. 14)

Biológia29: 565. 1974.

Basionimo: *Tetrastrum anomalum* G.M. Smith, Transactions of the American Microscopical Society 45: 187,pl. 15, fig. 21-27. 1926.

Cenobia flat, formed by 2-4 cells arranged in a pair on top of each other; cells cylindrical-arched, poles rounded, somewhat truncated, large spines, as long as the cell, joined in pairs by convex sides; cell length 5.0-11.0 pm, width 2.0-3.1 pm, 1-2 polar spines, straight, divergent, 4-10 pm long; parietal chloroplastidium, 1 pyrenoid.

Geographical distribution in the state of Sao Paulo

IN LITERATURE: **Municipality of Sao José dos Campos** [CARDOSO 1979: 79, fig. 100-102, as *Scenedesmus anómalas* (G.M. Smith) Tiffany var. *anómalas*]. **Municipio de Sao Paulo** (SANT'ANNAETAL. 1989: 95, fig. 60).

MATERIAL EXAMINED: only material from the literature.

Comment

Didymogenes anómala (G.M. Smith) Hindák has spines, a feature that differentiates this species from *D. palatina* Schmidle, which does not. The specimens shown in SANT'ANNA *ET AL.* (1989) have the same characteristics as the specimens in KOMÁREK & FOTT (1983), however, the specimens in SANT'ANNA *ET AL.* (1989) had slightly smaller measurements than those in KOMÁREK & FOTT (1983), which were around 6.0-17.8 pm in length and 1.2-5.0 pm in width. According to HINDÁK (1974), there is certainly great variability in the length of cell spines and this characteristic can differentiate species. The spines may be different from those in *Didymogenes,* where they are more delicate, thinner and gradually taper from the base. HORTOBÁGYI (1948) proposed several varieties, such as var. *acaudatus* without spines (= *D. palatina,* HORTOBÁGYI 1948: pl. 1, fig. 5), var. *bicaudatus* with two opposite spines on the cenobium (HORTOBÁGYI 1948: pl. 1, fig. 7) and var. *tetraspinosus,* where each cell of the cenobium has two spines on each pole (HORTOBÁGYI 1948: pl. 1, fig. 8). This procedure was based on the classical and exclusively morphological taxonomy of the genus *Scenedesmus,* where the number of spines represents a taxonomically significant characteristic. However, SMITH (1926) did not give much taxonomic weight to the characteristic of subpolar spines, stating that they were much less common. At the same time, he also reported that in the case of eight-celled syncenobium, the inner cells generally lack spines (SMITH 1926). BOURRELLY (1962) also mentioned that the number of spines in species is very variable, but stated that supernumerary spines (double in number) or their absence may occur only sporadically. UHERKOVICH (1968) observed, in agreement with SMITH (1926), that it is relatively rare for spines to occur only on the outer cells of an eight-celled syncytium (HORTOBÁGYI 1948:pl. 1, fig.9).

According to HINDÁK (1974), the number of spines may be a taxonomic characteristic in varieties of great importance; however, he suggests that further observations are still needed to corroborate this assumption. On the other hand, the author believes that it is extremely plausible that *D. palatina* Schmidle and *D. anómala* (G.M. Smith) Hindák are one and the same species, where the number of spines and their length or absence are not significant for infra-specific taxonomy if these characteristics do not combine with other differentially important characteristics. However, these latter characteristics have not yet been found. HINDÁK (1974) believes that pure cultures of material from this species may prove the permanence of some morphological characteristics.

Didymogenes palatina Schmidle (Fig. 15)

Hedwigia45: 35, fig. 1-4. 1905.

cenobia flat, formed by 4 cells, cenobia arranged one pair on top of the other; cells reniform, joined in pairs by convex sides, poles rounded, without spines; cell length 7.0-12.0 pm, width 2.0-3.0 pm; parietal chloroplastid, 1 pyrenoid.

Geographical distribution in the State of Sao Paulo

IN LITERATURE: **Municipio de Sao Paulo** (SANT'ANNAETAL. 1989: 95, fig. 57). EXAMINED MATERIAL: only material from literature.

Comment

Didymogenes palatina Schmidle is sufficiently distinct from *D. anómala* (G.M. Smith) Hindák by the absence of spines on its cells. It is known that loss or gain of spines can occur in older cells. A potential confusion exists, however, due to the fact that the species has no other morphological characteristics that differentiate them. According to HINDÁK (1974), cells with spines were examined in 1973 in Lake Zamecky; on the other hand, cells with and also without spines were seen near Hrusov in the Danube River in Bratislava (HINDÁK 1974: pl. 2, fig. 5-7, pl. 3, fig. 5). Separating the two species above was, in this case, practically impossible. The proposal is therefore to prepare cultures of these various materials in order to prove that the absence of spines is a permanent and unique feature for distinguishing the two species: D. *palatina* Schmidle and D. *anómala* (G.M. Smith) Hindák.

We now include the illustration by KOMÁREK & FOTT (1983) due to the lack of good representations in Brazilian literature.

Tetrachlorella Korsikov 1953

Cenobia formed by 4 cells arranged in a particular way, arranged in a single plane and surrounded by mucilage. The cells are ellipsoidal and are distributed in 2 pairs whose longitudinal axes are parallel to each other. The inner cells join more or less parallel to each other or slightly converge, half inclined, outer cells join obliquely to both ends of the inner cells, from where the cenobium appearance is alternated. Cell wall smooth or with irregular warts located at the poles. Reproduction by 4 autospores. Parietal chloroplastidia with 1 pyrenoid (COMAS 1996).

Only one species identified:

***Tetrachlorella alternans* (G.M. Smith) Korsikov** (Fig. 16)

Uchenyezapiski Gorkovskogo gosudarstvennogo universiteta9: 116-118, 126. 1939.

Basionimo: *Crucigenia alternans* G.M. Smith, Transactions of the American Microscopical Society45: 185, pl. 14, fig. 14-18. 1926.

Cenobia flat, formed by 4 cells, cells arranged alternately; cells oblong, irregular, not completely joined together, cells slightly inclined; cell wall smooth; cell length 5.8-10.7 pm, width 2.7-6.7 pm; parietal chloroplastid, 1 pyrenoid.

Geographical distribution in the state of Sao Paulo

IN THE LITERATURE: None. First mention of the occurrence of the species in the state of Sao Paulo.

EXAMINED MATERIAL: **Municipality of Cananéia**, Ilha Comprida, 120 m from the sea, lagoon, 7-111-1975, *D.M. Vital* (SP130813). **Municipality of Mirante do Paranapanema**, SP- 272, km 30.5, acude, metafiton, 16-V-2001, *C.E.M. Bicudo &D.C. Bicudo,* 22°16'41.0"S, 51°48'16.5"W, conductivity 40 pS.cm^{-1} , pH 6.0 (SP355396). **Municipality of Pirassununga**, SP-225, km 23, Bairro Cascalho, lake, ?-XII-1973, *P.A.C. Sema* (SP123900). **Municipality of Ribeirao Branco**, no precise location, 19-V-1972, *D.M. Vital* (SP130427). **Municipality of Sao Paulo**, Parque Estadual das Fontes do Ipiranga, Lago das Ninféias, 03- VIII-2007, *T.R. Santos,* 23°38'08"S and 23°40'18"S, 46°36'48"W and 46°38'00"W (SP399783).

Comment

This species is made up of four irregular, alternating cells that are not completely attached to each other. The cells are slightly inclined and have a smooth cell wall. According to COMAS (1996), the cenobia of *Tetrachlorella* have a particular arrangement, as they are not flat as in the other members of the subfamily and the cells are not completely joined to each other. The formation of cell groups composed of four irregularly alternating and more or less joined cells could place *Tetrachlorella* in the subfamily Scenedesmoideae or Crucigenoideae, but it is traditionally classified in the latter. According to KOMÁREK & FOTT (1983), during reproduction the autospores are not released by the rupture of the maternal cell wall but by the gelatinization of the wall, which could bring *Tetrachlorella* closer to the Oocystaceae.

Representatives of *Tetrachlorella alternans* (G.M. Smith) Korsikov were found in four localities in the state of Sao Paulo. In these four localities, the species has always formed populations with many individuals, so there is no difficulty in identifying it thanks to the characteristic arrangement of the cells in its cenobiums.

Tetrastrum Chodat 1895

More or less square flat cenobiums, sometimes with 1 small central spine, composed of 4 crossed cells, sometimes with a mucilaginous envelope, which can form transitional syncenobiums. Cells more or less triangular, with rounded or convex outer margins, in some species with 1-7 spines oriented more or less in the plane of the cenobium. Chloroplastidium parietal, with or without pyrenoid. Cell wall smooth or with warts or spines. Reproduction by 4-8 autospores oriented in the same direction as the maternal cenobium, released by the rupture of the maternal cell wall (COMAS 1996).

Key to identifying the species studied:
1. Cells with thorns2
1. Cells without thorns3
   2. Cells with only 1 thorn .... *T. elegans*
   3. Cells with more than 1 ....thorn4
4. Pyrenoid present..*T. triangulare*
5. Pyrenoid absent..*T. komarekii*
   6. Cells with 2 spines of different sizes..*T. heteracanthum*
   7. Cells with 2 to 5 equally sized spines...*T.   staurogeniaeforme*

Tetrastrum elegans Playfair (Fig. 17)
Proceedings of the Linnean Society of New South Wales 41: 832. 1917.

Cenobia flat, formed by 4 cells, cenobia arranged crosswise, with 1 small spine in the center; cells subtriangular, outer margins rounded, 1 long spine; cell length 5.0-7.6 pm, spine length 13.0-20.0 pm; parietal chloroplastid, 1 pyrenoid.

Geographical distribution in the State of Sao Paulo

IN LITERATURE: **Municipalities of Sao Bernardo do Campo** and **Sao Paulo** (SANT'ANNA 1984: 260, fig. 176). **Municipality of Sao Paulo** (SANT'ANNA *ETAL.* 1989: 97, fig. 99).

MATERIAL EXAMINED: only material from the literature.

Comment

Tetrastrum and *Crucigenia* are genera that are very similar to each other and the characteristics that separate them are, according to NOGUEIRA (1991), the shape of the intercellular spine and the presence or absence of ornamentation in *Tetrastrum*. According to SANT'ANNA (1984), *Tetrastrum elegans* Playfair is a species that is not very well delimited, being very similar to *Tetrastrum staurogeniaeforme* Lemmermann and *Tetrastrum triacanthum* Korsikov due to the great variability in the thickness, number and distribution of the spines, which means that these characteristics should not be used in isolation to delimit these species. According to KOMÁREK (1974), there is regularity in the distribution and shape of these spines. Thus, in *Tetrastrum staurogeniaeforme* Lemmermann the spines are thicker at the base, while in *Tetrastrum elegans* Playfair they are equally delicate throughout. ECHENIQUE *ET al.* (2004) presented specimens with features very similar to those observed by SANT'ANNA *ET AL.* (1989), i.e. with cenobia arranged crosswise leaving a small space in the center of the cenobium and subtriangular cells. KOMÁREK (1983) studied only one population, in which the majority of the cenobia found did not have spines, corresponding morphologically to *Tetrastrum glabrum* (Roll) Ahlstrom & Tiffany. However, several specimens were also found with solitary spines on the cells, which were identified as *T elegans* Playfair.

Tetrastrum heteracanthum (Nordstedt) Chodat (Fig. 18-20)
Bulletin de l'Herbier Boissier 3: 113. 1895.

Basiönimo: *Staurogenia heteracantha* Nordstedt. Botaniska Notiser 1882: 56. fig. text AB. 1882.

Flat cenobium. formed by 4 cells arranged crosswise. with 1 small square space in the center; rounded cells. outer margins with 2 spines of unequal size. straight or not. inner margins almost straight; diameter 3.0-7.62 pm. length of spines 3.0-10.8 pm; parietal chloroplastid. 1 pyrenoid.

Geographical distribution in the State of Sao Paulo

IN LITERATURE: **Municipalities of Atibaia** and **Santo André** (SANT'ANNA 1984: 261. fig. 177). **Municipality of Ribeirão Preto** (SILVA 1999: 292. fig. 73). **Municipality of Sao Paulo** (SANT'*ANNAETAL.* 1989: 97. fig. 100-101).

EXAMINED MATERIAL: **Municipality of Guaratinguetá**. Club of 500. lake. 1- IV-1966. *C.E.M. Bicudo* (SP96965). **Municipality of Limeira**. SP-151. between km 3 and 4. acude. 05-V-2000.

C.E.M. Bicudo & S.P. Schetty (SP365687). **Municipality of São Paulo**. Ipiranga Fountains State Park. Garbas Lake. 17-I-2007. *M. Borduqui*. 23° 38'08"S and 23° 40'18"S. 46O36'48"W and 46O38'00"W (SP399782).

Comment

Rounded cells that leave a small square space in the center of the cenobium. Each cell has two spines, one of which is much larger than the other. *Tetrastrum heteracanthum* (Nordstedt) Chodat is easily confused with *Tetrastrum homoiacanthum* (Huber-Pestalozzi) Comas which. according to COMAS (1996). differs from *Tetrastrum heteracanthum* (Nordstedt) Chodat because its spines are the same size. this is a stable characteristic in *T. homoiacanthum* (Huber-Pestalozzi) Comas. COMAS (1984) also stated that the presence, size and number of spines are important characters within the genus, which is why he considered *T. homoiacanthum* (Huber-Pestalozzi) Comas to be an independent species. SANT'ANNA *ET AL.* (1989) and SILVA (1999) examined specimens that were absolutely identical in morphology to those described in KOMÁREK & FOTT (1983), but with a minimal metric difference: the current specimens from the state of São Paulo are smaller than those recorded in KOMÁREK & FOTT (1983). According to SANT'ANNA (1984), *T. heteracanthum* (Nordstedt) Chodat is easily identifiable by the presence of two spines of unequal size per cell and the eccentric arrangement of these spines on the free margin of the cells. AHLSTROM & TIFFANY (1934) mentioned that the small spine in this species may not appear in any or all of the cells of the cenobium and that the species may therefore take the form of an *'elegans\'*. The larger spine varies greatly in length and thickness. The same individual may have more fragile thorns and others that are more resistant. Also in the specimens observed by AHLSTROM & TIFFANY (1934), a certain uniformity was seen with regard to the curvature of the spines, since they always occur straight and, with very few exceptions, curved. Finally, the small thorn usually alternates with the long thorn, but this arrangement can also vary.

According to KOMÁREK (1974), the planes of division of the mother cell shown in the illustration correspond to the initial stages of protoplast differentiation and not to the position of the young cenobia before their release from the mother cell.

Representatives of this species were recorded in three localities in the state of Sao Paulo and in all three, populations made up of many individuals were found. A large majority of very typical individuals with rounded cells were observed, alongside a few others with almost triangular cells. The spines also showed variation in the degree of their curvature, which was not always straight as in the typical specimens of this species. Some specimens had curved spines, both the largest and the smallest in the cell, and this curvature could be very pronounced. This feature is also not very common within the species, as the specimens considered typical have completely straight spines. *Tetrastrum heteracanthum* (Nordstedt) Chodat.

Tetrastrum komarekii Hindák (Fig. 21)

Biologické Práce 23(4): 164. 1977.

Cenobium flat, formed by 4 cells, cells arranged crosswise leaving 1 small central spathe, sometimes absent, with or without mucilaginous envelope; cells subtriangular to trapezoidal, outer margins slightly convex, rounded, cell length 10-12 qm, width 3.0-6.0 qm; chloroplastid parietal, without pyrenoid.

Geographical distribution in the state of Sao Paulo

IN LITERATURE: **Municipio de Sao Paulo** (Tuca 2002: 244, fig. 42, *TucaETAL.* 2006: 165, fig. 53).

MATERIAL EXAMINED: only material from the literature.

Comment

Tetrastrum komarekii Hindák is typical for its more or less square cenobiums made up of four cells, with a central spire with a not very well-defined shape, subtriangular cells and a cell wall that is either smooth or granular. According to COMAS (1996), part of the previous identifications of *Tetrastrum triangulare* (Chodat) Komárek should be revised, as it is *T. komarekii* Hindák.

However, the existence of *T. komarekii* Hindák can be questioned, since the only difference between this species and *T. triangulare* (Chodat) Komárek is the absence of a pyrenoid in the former. According to COMAS (1996), results from molecular biology, the presence or absence of pyrenoid has not universally served to differentiate species. Until proven otherwise, however, *T.komarekii* Hindák should be retained.

Tetrastrum staurogeniaeforme (Schroder) Lemmermann (Fig. 22)

Berichte derdeutschen botanische Geseslchaft 18: 95. 1900.

Basionimo: *Cohniella staurogeniaeformis* **Schroder,** Berichte der deutschen botanische Geseslchaft 15: 373. 1897.

Cenobium flat, formed by 4 cells, cells arranged in a cross, with intercellular spire; cells subtriangular-rounded, outer margins semi-circular, sometimes flattened, inner part of cell with rectangular margin, 2-5 short or long spines; cell length 5.2-5.8 pm, cell width 4.8-6.0 pm, polar spines 10-20 pm; parietal chloroplastidium, 1 pyrenoid.

Geographical distribution in the State of Sao Paulo

IN LITERATURE: **Municipality of Santo André** (SANT'ANNA 1984: 265, fig. 179180).

MATERIAL EXAMINED: only material from the literature.

Comment

According to SANT'ANNA (1984), *Tetrastrum staurogeniaeforme* (Schroder) Lemmermann is typical in that it has two to five polar spines of approximately equal size arranged on the free margin of the cells. According to AHLSTROM & TIFFANY (1934), the shape of *T. staurogeniaeforme* (Schróder) Lemmermann is typically longer than wide and the four cells of the cenobium do not touch in the center of the cenobium. Most of the specimens studied by the various authors showed this typical shape, which was conveniently called cruciform. The authors also state that individuals with few spines have longer than wide shapes and that the four cells touch in the center of the cenobium. This form of cenobium is called axial. Generally, the shape of the cenobium of *T. staurogeniaeforme* (Schróder) Lemmermann is cruciform and that of *T. heteracanthum* (Nordstedt) Chodat is axial, but individuals of T. *staurogeniaeforme* (Schróder) Lemmermann may have an axial arrangement and those of *T. heteracanthum* (Nordstedt) Chodat a cruciform arrangement. BICUDO & VENTRICE (1968) described *T. staurogeniaeforme* (Schróder) Lemmermann f. *brasiliense* C. Bicudo & Ventrice and mentioned that this taxonomic form differs from the typical one due to the much larger size of the cells, which reached twice the maximum limits proposed by AHLSTROM & TIFFANY (1934). BICUDO & VENTRICE (1968) believed, however, that this was a simple morphological variation within the species. KOMÁREK (1974) stated that there is great variation in the length of the spines, as there are popular ones with uniformly small or variably long spines. The taxonomic value of these modifications (var. *longispinum*, f. *crassispinosus*, f *crassispinum*, f *obtusum*, etc.) is unclear. The same author also mentioned that even when *T. staurogeniaeforme* (Schróder) Lemmermann loses its spines under some circumstances, the validity of the species with a smaller amount of these appendages should not be rejected.

AHLSTROM & TIFFANY (1934) and KOMÁREK (1974) mentioned the possibility of uniting *T. staurogeniaeforme* (Schróder) Lemmermann, *T. heteracanthum* (Nordstedt) Chodat, T. *elegans* Playfair and *T. glabrum* (Roll) Ahlstrom & Tiffany, as they observed that, although we are talking about distinct species, they have a very variable number of spines. They concluded that even when *T. staurogeniaeforme* (Schróder) Lemmermann completely loses its spines or they are reduced in number, there is still doubt as to whether the species remains independent of the others.

Tetrastrum triangulare **(Chodat) Komárek** (Fig. 23)

Archiv fur Protistenkunde 116: 30, fig. 8. 1974.

Cenobium flat, formed by 4 cells, more or less square, with 1 small space in the center; triangular cells, arranged crosswise, outer margins almost straight, sometimes slightly rounded; cell length 2.51-5.5 pm, diameter 2.4-4.56 pm; parietal chloroplastidium, 1 pyrenoid.

Geographical distribution in the state of Sao Paulo

IN LITERATURE: **Municipality of Sao Paulo** (FERRAGUT *ET AL.* 2005: 156, fig. 101; SANT'ANNA *ET AL.* 1989: 97, fig. 104).

EXAMINED MATERIAL: **Municipality of Guaratinguetá**, Clube dos 500, lake, 01-IV-1966, *C.E.M. Bicudo* (SP96965). **Municipality of Juquiá**, BR-116, km 165, lake, 01-III- 1973, *C.EM. Bicudo & L. Sormus* (SP113664). **Municipality of Macedonia**, highway Alberto Faria, direction Mia Estrela-Macedonia, 2 km before the entrance to Macedonia, dam, 25-IV- 2001, *C.E.M. Bicudo, D.L. Costa & S.M.M. Faustino,* 20° 08'19,5"S, 50°11'56,4"W, conductivity 70 pS cm^{-1} , pH 6,6 (SP355366).

Comment

According to NOGUEIRA (1991), *Tetrastrum triangulare* (Chodat) Komárek is similar to *Westella botryoides* (W. West) De-Wildemann due to the sub-spherical shape of the cells. However, the arrangement of the daughter cenobia separates the latter from the former species. *Tetrastrum triangulare* (Chodat) Komárek resembles, in our opinion, *Crucigenia tetrapedia* (Kirchner) West & West. These two species are easily confused with each other. What separates them is the fact that the former has a very small square space in the center of the cenobium and the latter does not, as well as the production of the autospores, which in *Tetrastrum triangulare* (Chodat) Komárek is done according to a division plane forming angles and in *Crucigenia tetrapedia* (Kirchner) West & West in the shape of a cross. The specimens in SANT'ANNA ETAL. (1989) and FERRAGUT ET *AL.* (2005) show the same characteristics as those presented in KOMÁREK & FOTT (1983), which are quadrangular cenobia with four cells arranged crosswise, a rectangular central space and triangular-shaped cells. KOMÁREK (1983) observed in the samples collected in Cuba specimens without a pyrenoid and without an opening in the center of the cenobium, but which in cultivation showed a pyrenoid and a small rectangular opening in the center of the cenobium. HINDÁK (1977) divided the two types into two species, proposing the new species *T. komarekii* Hindák. for the populations without pyrenoids.

Tetrastrum triangulare (Chodat) Komárek was collected in three municipalities, forming populations with large numbers of individuals. No difficulty was encountered in identifying the representatives from the state of São Paulo, since the specimens observed always showed a small opening in the center of the cenobium.

Table 3. Differential characteristics between *T. elegans, T. heteracanthum, T. komarekii, T. staurogeaniaeforme* and *T. triangulare.*

Species	Pyrenoid	Thorns in the cells	Central opening of the cenobium	Cell wall
T. elegans (Sant'Anna, 1984)	present	1 long thorn	small rectangular space	smooth, with thorns
T. heteracanthum	present	2 unequally long	small square space	smooth, with thorns
T. komarekii (Tucci *etal.* 2006)	absent	Absent	small, sometimes absent	smooth
T. staurogeniaeforme (Sant'Anna, 1984)	present	2 to 5 equally long and regularly arranged	absent	smooth, with thorns
T. triangulare	present	Absent	small square space	smooth

Westella De-Wildeman 1897

Cenobia composed of crossed cells, elongated in one direction, usually forming syncenobia, where the spatially arranged daughter cenobia are joined by remnants of the maternal walls. The cell wall is smooth. Chloroplastidia parietal, with a pyrenoid (COMAS 1996).

Only one species identified:

***Westella botryoides* (W. West) De-Wildemann** (Fig. 24-26)

Bulletin de l'Herbier Boissier 5: 532. 1897.

Basionimo: *Tetracoccus botryoides* W. West, Journal of the Royal Microscopical Society 2: 735.1892.

Cenobia formed by 4 or 8 cells, forming syncenobia of up to 8 cells; spherical cells; cell diameter 4.0-10.0 pm; parietal chloroplast, 1 pyrenoid.

Geographical distribution in the State of Sao Paulo

IN LITERATURE: **Municipalities of Arujá, Cananéia, Rio Claro, Pirassununga** and **Santo André** (SANT'ANNA 1984: 267, fig. 181-183). **Municipality of Luiz Antonio** (SCHWARZBOLD 1992: 112). **Municipality of Sao Paulo** (SANT'*AnnaetAL*. 1989: 98, fig. 105).

EXAMINED MATERIAL: **Municipality of Ubatuba**, without precise indication of location, 29-I-1966, *O. Montes & R.R. Martins* (SP96890). **Municipality of Pirassununga**, SP-225, km 23, Bairro Cascalho, lake, ?-XII-1973, *P.A.C. Senna* (SP123900). **Municipality of Monte Alto**, highway between Monte Alto and Vista Alegre, lake, with grasses and *Typha,* 20-II-1992, *L.H.Z. Branco* (SP239233). **Municipio de General Salgado**, rodovia Jesulino da Costa Frota (vicinal road), 1.5 km from road SP-310, unspecified location, with Cyperaceae, grasses and *Typha,* 05-XII-1991, *L.H.Z. Branco* (SP239241).

Comment

According to SANT'ANNA (1984), *Westella botryoides* (W. West) De-Wildemann is characterized by the arrangement of the cells in groups of four, which can be isolated or connected to each other by remnants of the mother cell wall, forming multiple cenobia. The same author also mentions that this type of cell arrangement resembles that of *Dictyosphaerium pulchellum* Wood, however, the cells with flat faces in contact are characteristic of *Westella* and, according to PICELLI-VICENTIM (1987), the wall remnants that connect the cells of *Dictyosphaerium to each other are* markedly in the form of dichotomously branched threads. According to COMAS (1984), in samples he examined of material from Cuba, populations were observed in which the cells of the cenobiums are strongly joined together, as in *Tetrastrum.* However, in other populations, the cells appeared less close together, with small gaps between them. COMAS (1984) also mentioned that this type probably represents a taxonomic novelty without, however, defining whether it would be a species or a variety. The author also mentioned the existence of syncenobia which can vary from 16 to 100 cells.

Representatives of this species were found in four municipalities in the state of São Paulo. We found that *Westella botryoides* (W. West) De-Wildemann is well represented in the state. Large populations were found with many typical individuals representing the species, which was considered, in our opinion, easy to identify despite its similarity to material from *Dictyosphaeriumpulchellum* Wood.

Willea Schmidle 1900

Flat cenobiums, usually surrounded by remnants of the maternal cell wall, made up of 2-4 paired cells. The longitudinal axis of the cenobium passes through pairs of cells which are sometimes surrounded by mucilage. Oval or cylindrical cells. Thick cell wall, with smooth distal ends. Reproduction by 2-4 autospores which are, in the first case, oriented more or less perpendicularly to the longitudinal axis of the maternal cenobium and, in the second, oriented in the same direction as the maternal cenobium. Parietal chloroplastid with or without pyrenoid. All known species form syncenobia (COMAS 1996).

Key to identifying the species studied:
1. Cells oblong to oval, with pyrenoid *W.* .... *irregularis*
1. Elliptical oval cells, without pyrenoid *W.* .... *vilhelmii*

Willea irregularis (Wille) Schmidle (Fig. 27)
Berichte derdeutschen botanische Geseslchaft 18: 157. 1900.
Basionimo: *Crucigenia irregularis* Wille, Biologische Centralblat 18: 302. 1898.

Cenobia flat, formed by 4 cells in contact by the poles and lateral walls, with 1 small quadrangular central space in the cenobium, always forming compound cenobia; cells oblong to oval; cell length 8.0-12.0 pm, width 4.0-7.0 pm; parietal chloroplastid, 1 pyrenoid.

Geographical distribution in the state of Sao Paulo

IN LITERATURE: **Municipalities of Pindamonhangaba** and **Sao Paulo** (SANT'ANNA 1984: 269, fig. 184-187).

MATERIAL EXAMINED: only material from the literature.

Comment

Willea irregularis (Wille) Schmidle has irregularly grouped cenobia and always forms compound cenobia. According to KOMÁREK (1974), in *Willea* the major axis of the cenobium is between two cells. The author noted that the phytoplankton specimens had a smaller number of cells in the cenobium and a slight thickening at the poles of the cell wall, i.e. they were very different from the benthic materials, which had a greater number of cells and lacked the polar wall thickening.

SANT'ANNA (1984) studied several samples which showed, in addition to autospore reproduction, a spherical structure with a thickened wall and dividing plasma content. This suggested to the author that this structure could be a zygote. However, more studies are needed to reach any conclusions, as all the

genera belonging to the family have so far only reproduced by autospores.
***Willea vilhelmii* (Fott) Komárek** (Fig. 28)
Archiv fur Protistenkunde 116: 42. 1974.
Basionimo: *Dispora vilhelmii* Fott, Bulletin de l'Institut et du Jardin Botanique de l'Université de
Belgrade 2(3): 168. 1933.

Flat cenobiums, formed by 4-16 cells around the wall of the mother cell; oval-elliptical cells, sometimes with an irregular number of cells, a successive formation of daughter cenobiums, rhomboidal central space, cell length 4.55-5.05 pm, diameter 8.05-8.75 pm, parietal chloroplastidium, no pyrenoid.

Geographical distribution in the State of Sao Paulo

IN THE LITERATURE: nothing. First mention of the occurrence of the species in the state of Sao Paulo.

EXAMINED MATERIAL: **Municipality of Pindamonhangaba,** BR-116, km 270271, lagoon, 21-V-1966, *C.E.M.Bicudo* (SP96949).

Comment

According to KOMÁREK (1974), the original description of *Willea vilhelmii* (Fott) Komárek does not include any information about the pyrenoid. However, HINDÁK (1969) studied material from the species itself in cultivation and verified the presence of the pyrenoid. The specimens from the state of Sao Paulo also lacked pyrenoid. In our opinion, *Willea vilhelmii* (Fott) Komárek requires more in-depth studies to confirm the existence or absence of pyrenoids in the species.

Representatives of this species were only collected in Pindamonhangaba. In addition, few individuals representing this species were found in the only sample studied. *Willea vilhelmii* (Fott) Komárek is not a very common species in the waters of the state of São Paulo.

Table 4. Differential characteristics between *W. irregularis* and *W. vilhelmii.*

Species	Pyrenoid	Cell shape	Central opening of the cenobium	Number of cells per cenobium
W. irregularis (Sant'Anna, 1984)	present	oblong to oval	small, square space	4-100
W *vilhelmii*	absent	oval-elliptical	rhombus space	4-16

Subfamily Danubioideae *sensu* Komárek & Fott 1983

Cenobia with 4 cells touching at one end. The tetrads are more or less regular and can be joined by remnants of the maternal walls to form syncytia. Chloroplastidium parietal, with or without pyrenoid. Smooth or warty cell wall. Reproduction by 4-8 autospores united in autocoenobes of 4 cells each, which are released by rupture of the maternal wall.

Key to identifying the genera studied:
1. Syncenobes with up to 16 cells, with irregularities in the arrangement of the cells; cells touching or separated from each other, with encrustations on the cell surface ....*Pseudotetrastrum*
1. Cenobium formed by 4 cells arranged in a cross, with 1 quadrangular to rectangular spathe in the center of the cenobium; reniform cells, with the convex margin always facing the periphery of the cenobium .... *Tetranephris*

Pseudotetrastrum Hindák 1977

Free-floating cenobia formed by 4 cells, more or less flat, with the cells arranged in a square, forming syncenobia of up to 16 cells, with irregularities in the arrangement of the cells (spatially ordered multicellular groups appear secondarily), devoid of a mucilaginous envelope. Cells touching or separated from each other, with encrustations on the cell surface. Chloroplastidium parietal, with pyrenoid (KOMÁREK & FOTT 1983).

Only one species identified:

***Pseudotetrastrumpunctatum* (Schmidle) Hindák** (Fig. 29)

Biologické Práce 23(4): 142, pl. 57, fig. 1-17. 1977.

Basionimo: *Staurogenia multiseta* Schmidle var. *punctata* Schmidle, Bericht der deutsche botanische
Gesellschaft 18: 157, pl. 65, fig. 13-14. 1900.

Nomenclatural synonym: *Tetrastrum punctatum* (Schmidle) Ahlstrom & Tiffany, Bericht der deutsche
botanische Gesellschaft 18: 157, pl. 65, fig. 13-14. 1900.

Cenobia flat, square or slightly elongated, formed by 4 cells joined laterally, with spagos in the center of the cenobium; cells irregularly oblong, outer margins slightly convex; cell wall composed of warts derived from iron impregnations; cell length 1.9-6.95 pm, width 3.2-7.5 pm; parietal chloroplastidium, 1 pyrenoid.

Geographical distribution in the state of Sao Paulo

IN LITERATURE: **Municipio de Sao Paulo** [Sant'Anna *et al.* 1989: 97, fig. 102, as *Tetrastrumpunctatum* (Schmidle) Ahlstrom & Tiffany].

EXAMINED MATERIAL: **Municipality of Pindamonhangaba**, Fazenda Sao Joao, Sao Joao lagoon, 21-V-1966, *C.E.M. Bicudo* (SP96950). **Municipio de Santa Rita do Oeste**, vicinal, 3 km after the entrance to the town, stream, metafiton, 25-IV-2001, *C.E.M. Bicudo, D.L. Costa & S.M.M. Faustino,* 20°07'36"S, 50°48'0.9"W, conductivity 110 pS cm^{-1} , pH 6.8 (SP355394). **Municipio de Sao Paulo**, Parque Estadual das Fontes do Ipiranga, Lago das Garbas, 17-I-2007, M. *Borduqui,* 23°38'08"S and 23°40'18"S, 46°36'48'W and 46°38'00'W (SP399782).

Comment

SANT'ANNA ET AL. (1989) identified material from the Municipality of Sao Paulo as *Tetrastrum punctatum* (Schmidle) Ahlstrom & Tiffany, a name that is now considered synonymous with *Pseudotetrastrum punctatum* Hindák according to KOMÁREK & FOTT (1983). According to the latter author, *Pseudotetrastrum punctatum* Hindák has a 4-celled cenobium, made up of elliptical cells with dots (warts) at the poles, as well as the formation of a small square space in the central region of the cenobium.

According to HINDÁK (1977), *Tetrastrum punctatum* (Schmidle) Ahlstrom & Tiffany was transferred to a separate genus because it has a granular cell wall on the surface, a feature not found in *Tetrastrum* species. Another distinguishing feature between these two genera is the free, spherical-shaped cells of *Pseudotetrastrum*. However, the author above mentioned that this granulation can be variable, both in shape and location. These granules are usually observed free near the cell margins and, less frequently, on the surface. According to AHLSTROM & TIFFANY (1934), *Pseudotetrastrum punctatum* (Schmidle) Hindák seems to be a valid species. Gilbert M. Smith observed a punctate form of *Crucigenia quadrata* Morren which he considered identical to *Pseudotetrastrum punctatum* (Schmidle) Hindák var. *schmidlei* Schmidle. These authors also observed the Hindák var. *punctata* of *Crucigenia quadrata* Morren in material from Lake Michigan. AHLSTROM & TIFFANY (1934) finally observed collections of planktonic algae sent to them by H. Skuja containing specimens not identical to the punctate form of *Crucigenia quadrata* Morren and therefore considered them to be distinct species.

Pseudotetrastrum punctatum (Schmidle) Hindák was found in three localities in the state of São Paulo, with only a few individuals representing the species in each of these three localities. The species had already been mentioned by SANT'ANNA ETTL. (1989) for the state of Sao Paulo as *Tetrastrum punctatum* (Schmidle) Ahlstrom & Tiffany and the material studied showed warts located on the cell margin, which is diagnostic of the species. The material examined did not present any difficulties for taxonomic identification, as it presented very typical characteristics and showed no morphological variation.

Tetranephris Leíte & C. Bicudo 1977

Colonial individuals who live freely in the system. The cenobium is made up of 4 cells arranged in a cross, with 1 quadrangular to rectangular spire in the center. Multiple cenobiums are common. The cells are reniform and the convex margin always faces the periphery of the cenobium and the poles are rounded to acuminate-rounded. Remnants of the mother cell wall may remain in the form of delicate filaments that join the cenobia together to form multiple cenobia. The chloroplastidium fills the cell internally and has no pyrenoid (BICUDO & MENEZES 2006).

Only one species identified:

***Tetranephris brasiliensis* Leíte & C. Bicudo** (Fig. 30-31)

Phycologia 16(3): 231, fig. 1-8. 1977.

Cenobia formed by 4 cells, cells arranged cruciately, with 1 square spire in the center of the cenobium; oval to reniform cells, convex margins; cell length 4.0-5.8 pm, width 6.0-8.6 pm; parietal chloroplastid, no pyrenoid.

Geographical distribution in the state of Sao Paulo

IN LITERATURE: **Municipality of Cananéia** (LEITE & BICUDO 1977: 231, fig. 1-8; SANT'ANNA 1984: 258, fig. 172-175). **Municipality of Juquiá** (SANT'ANNA ET AL. 1988: 91, fig. 44-45).

EXAMINED MATERIAL: **Municipality of Santo André**, Estacao Biológica do Alto da Serra de Paranapiacaba, stream near the headquarters, 20-I-1966, *C.E.M. Bicudo* (SP96922). **Municipality of Itanhaém**, SP-55, km 332.7, pond, with *Typha* and *Eichornia,* 28-II-1990, *L.H.Z. Branco* (SP188434). **Municipality of Guará**, Pioneiros District, SP-330, km 393.25 highway between Guará and Sao Joaquim da Barra, near Sao Joaquim da Barra, on the right towards Guará-Sao Joaquim, pond with Cyperaceae and Poacae, 02-IX-1990, A.A.J. de Castro (SP255739).

Comment

LEITE & BICUDO (1977) proposed the genus *Tetranephris from* material collected on Ilha Comprida, Municipality of Cananéia, southern region of the State of Sao Paulo. *Tetranephris brasiliensis* Leite & C. Bicudo presents cenobia formed by four oval to reniform cells that are arranged crosswise, leaving a square spine in the center of the cenobium.

According to SANT'ANNA (1984), *Tetranephris* was a monospecific genus until a second species, *Tetranephris europaea* (Hindák) Komárek, was included in it. *Tetranephris europaea* (Hindák) Komárek is also a rare species, having only been recorded from Slovakia and Hungary. This last species has slightly smaller cells than *T. brasiliensis* Leite & C. Bicudo and a slight difference in cell shape (not so reniform). HEGEWALD ET AL. (2002) also identified *T. brasiliensis* Leite & C. Bicudo, however, the specimens they examined had non-curved cells and, above all, were very close together. The material from *T. brasiliensis* Leite & C. Bicudo studied by COMAS (1984) differs from the standard material of the species in that the cells are more arched and the apexes are wider. PICELLI-VICENTIM (1987) compared *T. brasiliensis* Leite & C. Bicudo with *Gyoerffyana humicola* Kol. & Chodat (endemic to Amazonia); however, the latter is different in that it has chloroplastids with pyrenoids and mucilage surrounding the cell groups. PICELLI-VICENTIM (1987) compared *T. brasiliensis* Leite & C. Bicudo with *Hofmania appendiculata* Chodat and differentiated them by the fact that the former had reniform cells, without any appendix on the free margin of the cells and did not have a pyrenoid. SANT'ANNA & MARTINS (1982) recorded, for the first time, the occurrence of the genus *Tetranephris* in the Amazon region, as it was then only known from the southeast of Brazil.

The specimens now identified are very similar to the originals in LEITE & BICUDO (1977). *Tetranephris brasiliensis* Leite & C. Bicudo was collected from three sites in the state of São Paulo, and large numbers of specimens were never popular in these samples. The specimens that occurred in these samples proved to be very typical of the species, and did not cause any difficulty in their taxonomic identification.

Subfamily Dimorphococcoideae *sensu* Komárek & Fott 1983

Cenobium with 4-8-16 cells generally situated in different planes; slightly or markedly alternating cells, outer and inner cells sometimes morphologically quite different, joined by their walls or appendages; parietal chloroplastidium, with pyrenoid; smooth cell wall, sometimes with apical thickenings.

Dimorphococcus A. Braun 1855

Cenobium formed by numerous groups of 4 cells. The cells that make up each of these groups are characteristically of 2 morphological types, i.e. the two outermost cells have a slightly different shape to the two cells located further inland in the gelatinous matrix. Reproduction by the formation of autospores. Chloroplastidia with 1 pyrenoid.

Only one species identified:

Dimorphococcus lunatus A. Braun (Fig. 32)

Algarium unicellularium genera nova vel minus cognita, pr^missis observationibus de algis unicellularibus in genere. 44. 1855.

Cenobia in several planes, formed by a very variable number of cells; outer cells cordiform to reniform, joined to the inner cells by the convex margin, inner cells oval-cylindrical, joined by the widest parts, in both cases rounded poles, may show thickening of the cell wall; cell length 3.35-13.7 pm, width 2.5-8.3 pm; parietal chloroplastidium, 1 pyrenoid.

Geographical distribution in the state of Sao Paulo

IN LITERATURE: **Municipality of Juquiá** (SANT'ANNA *et al.* 1988: 88, fig. 14). **Municipality**

of Luiz Antonio (PERES & SENNA 2000: 473, fig. 29; SCHWARZBOLD 1992: 112, fig. 6). **Municipality of Moji-Gua^u** (MARINHO 1994: 47, fig. 3). **Municipality of Sao Carlos** (HINO & TUNDISI 1977: 106, fig. 141).

EXAMINED MATERIAL: **Municipality of Guaratinguetá**, Clube dos 500, lake, 01- IV-1966, *C.E.M. Bicudo* (SP96965). **Municipality of Itu**, SP-280, km 77, lake, 11-V-1977, *C.R. Leite* (SP139733). **Municipality of Sao Paulo,** Parque Estadual das Fontes do Ipiranga, artificial lake, 5-VIII-1973, *C.R. Leite* (SP130453).

Comment

According to COMAS (1996), *Dimorphococcus lunatus* A. Braun has two morphological cell types: (7) reniform outer cells, with a thickened wall at the cell poles and (2) more or less cordiform outer cells, smaller than the reniform ones, with a wall with or without slight thickening at the cell poles. This second type is the one that appears most frequently in the tropics. Some specimens identified as *D. lunatus* A. Braun or *Dimorphococcus cordatus* Wolle are most likely representatives of *Dimorphococcus fritschii* (Crow) Jao. BICUDO & VENTRICE (1968) found individuals with cells from the same group that show dimorphism, i.e. two cells are approximately elliptical or subcylindrical and the other two cordate. HINO & TUNDISI (1977) agreed with BICUDO & VENTRICE (1968) and also mentioned that *Dimorphococcus lunatus* A. Braun had two types of cells in the same cenobium. MARINHO (1994) also found two cell types, but with slightly larger dimensions, reaching 15.0 pm in length and 9.0 pm in width. SANT'ANNA *ET al.* (1988) found colonies made up of several groups of four cells, connected by branched strands of mucilage, in which the two central cells are oblong and the two peripheral ones cordiform and reniform, also agreeing with the other authors in relation to cell dimorphism.

Dimorphococcus lunatus A. Braun was found in three locations in the state of Sâo Paulo and in all of them no populations consisting of large numbers of individuals were found. However, the species was considered easy to identify due to its cellular dimorphism. It is worth remembering that *Dimorphococcus lunatus* A. Braun can be confused with representatives of *Dictyosphaerium,* but the shape of the cells of the latter is rounded and very different from that of the species in question, in addition to never showing cell *dimorphism, and it* is also a species very similar to *Dimorphococcopsis* species.

Subfamily Desmodesmoideae *sensu* Komárek & Fott 1983

According to HEGEWALD & HANAGATA (2000), the subfamily Desmodesmoideae is made up of cells with more or less obtuse or truncated cell poles. Cell walls with sporopollenin structures. May have spines, ribs, denticles and warts and rosettes. Cells linearly or obliquely arranged.

Identification key for the genera studied:
1. Flat cenobium formed by 2-16 cells arranged linearly or alternately, forming 1 row whose cells may be arranged in 1 or 2 series: cell wall ornamented with sporopollenin and cellulose and with warts, spines, denticles or rosettes..*Desmodesmus*
1. Cenobium consists of only 2 cells arranged next to each other along their longest axes; cell wall with cellulose, decorated with granular inorganic material or warts ....*Pseudodidymocystis*

Desmodesmus (Chodat) An, Friedl & Hegewald 1999

Many subgenera were created for *Scenedesmus* and finally the following 3 were accepted by HEGEWALD (1978): *Acutodesmus, Scenedesmus* and *Desmodesmus.* The subgenus *Desmodesmus* was elevated to genus level by ANET AL. (1999), but these authors were unable to separate the subgenus *Acutodesmus* from *Scenedesmus.*

AN *ET AL.* (1999) transferred only 5 species of the former *Scenedesmus* to the new genus: *D. arthrodesmiformis* (Schroder) An *et al., D. bicellularis* (Komárek) An *et al., D. denticulatus (*Lagerheim) An et al., D. *lefevrei* Komárek and *D. serratas (*Corda) An *et al.* HEGEWALD (2000) transferred 55 taxa between species and varieties. To date, many of these taxa (generally based on morphological criteria) are conceived in a broad sense, but HEGEWALD (2000) recognizes and describes some varieties and species generally have broad lists of synonyms. Many of these names have been applied to morphotypes that actually exist and have been recorded by various authors.

Free-living colonial individuals. Flat cenobiums made up of 2, 4, 8 or 16 cells arranged linearly or alternately, with their longest axes parallel to each other. The cells can be ellipsoid or ovoid and all the same or of two types, i.e. the outer cells of the cenobium are of one type and the inner cells of another. The cell wall is decorated with ribs, warts, spines, denticles or rosettes. The spines can occur either on the poles of the outer cells or on the poles of the inner cells; there are no species without spines or without some kind of ornamentation. Parietal chloroplastidium, with 1 pyrenoid (BICUDO&MENEZES2006).

Tipos de espinho com arranjo equatorial		Tipo de espinhos com arranjo lateral-longitudinal				
2 espinhos equatoriais na parte mediana da célula externa	Espinhos equatoriais nas células externas e nas células centrais	Número de espinhos laterais 0-2, **geralmente 1**			Número de espinhos laterais 0-5, **geralmente 2-3**	
		Espinhos longos nas células centrais	Espinhos curtos nas células centrais	Espinhos curtos nas células centrais	Forma da célula oval elíptica	Forma da célula oblonga
			Com "costelas"	Sem costelas		
flavescens	*asymmetricus*	*pleiomorphus*	*spinosus*	*abundans*	*kissii*	*subspicatus*
Hegewald 1993	Hegewald & Schmidt 1989	Hegewald 1989	Komárek & Ludvík 1972, Hegewald et al. 1990	Komárek & Ludvík 1972, Hegewald & Schnepf 1991	Hegewald 1989	Komárek & Ludvík 1972

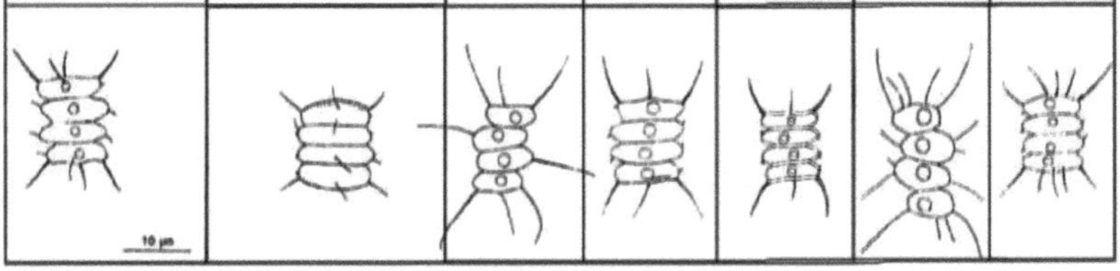

Figure 2 - Comparative table of species with types of spines and arrangements (HEGEWALD *ET al.* 2001).
Key for identifying the species and varieties studied:

1. Cells with denticles and polar spines and 1 lateral spine in the median part of the outer cell...*Desmodesmus* sp. 1
1. Cells without polar denticles, with polar spines and 1 lateral spine in the middle of the outer cell2
  2. Cells with granular margin ....*Desmodesmus* sp. 2
  3. Cells without granular margin3
4. Cells with v-shaped denticles ....*Desmodesmus* sp. 3
5. Cells without v-shaped denticles4
  6. With poorly developed spines, basically decorated with denticles, ribs or granules5
  7. With well-developed spines, especially at the poles of the outer cells13
8. Cells with denticles at the cell poles and longitudinal ribs from pole to pole or interrupted, sometimes with winged structures (combs)..*D. brasiliensis*
9. Cells with denticles only at the cell poles, without ribs, sometimes with accessory granules6
  10. 1 denticle at each cell pole; cell margins with 1 longitudinal series of rib-like denticles; ellipsoid-fusiform cells ....*D. serratus*
  11. 1-3 denticles at each cell pole, when present at the cell margins they are scarce and do not form rib-like series; cells of other shapes7
12. Cenobium with irregularly oblong cells, markedly alternate   *D. denticulatus* var. *denticulatus*
13. Cenobium with cells of other shapes, distributed more or less in a line 8
  14. Poles of the outer cells with 1 or more denticles oriented with respect to the cenobium9
  15. Poles of outer cells with 1 or more denticles without any orientation 10
16. 1-2 polar denticles (on the cell margins of the outer cells there can sometimes be 1 series of tiny structures resembling small teeth), 1 of them along the longitudinal axis of the cell, the other forming a right angle to the first; oval-fusiform cells, truncated poles.........................*D. spinulatus*
17. 1(-3) polar denticles, only 1 well-developed, oriented towards the inside of the cenobium, the others without a defined orientation; fUsiform-elliptical cells, poles obtuse or slightly truncated   *D. arthrodesmiformis*
  18. Cylindrical cells, polar denticles with a tiny appearance granules both at the poles and at the cell margins11
  19. Oblong cells, 1-2 polar denticles, sometimes accompanied by 1-2 denticles at the cell margins that do not look like granules13
20. Outer cells gently arched; poles rounded-truncated, 1-3 small, sometimes inconspicuous, granule-like polar denticles; granules sometimes present on cell margins.......................*D. lunatus*
21. Outer cells with straight or convex margins, not arched; poles rounded-obtuse or rounded-truncated, 1-3 conspicuous polar denticles; granules absent12
  22. Rounded-truncated cell poles, each with 1 -3 denticles......................... *D. aculeolatus*

23. Poles rounded, each with 1 denticle...................................*D. denticulatus* var. *linearis*
24. Cells with 4 well-developed main spines, 1 at each pole of the outer cells; if there are accompanying spines, they are located at the poles, never laterally; denticles or ribs may appear next to the spines14
25. Cells with the 4 main spines and lateral spines25
   26. Cells with only 4 main spines at the poles of the outer cells; exceptionally, accessory spines at the poles of the inner cells15
   27. Ribs, denticles and short spines can occur alongside the main spines 19
28. Elliptical-oblong cells, markedly alternate ....(D. *intermedius*) 16
29. More or less cylindrical or cylindrical-oval cells, more or less aligned20
   30. Bicaudate cells.. *D. intermedius* var. *acutispinus*
   31. Non-bicaudate cells .... *D. intermedius* var. *intermedius*
32. Perforated Cenobium (with intercellular spathes) .... *D.perforatus*
33. .... Non-perforated cenobiums18
   34. Cells joined together along the entire length of their wall;
   extem cells with 1 slight convexity in the middle;
   inner cells with broadly rounded poles, no spines
   accessories; cenobia usually 4-celled.. *D.   communis*
   35. Cells not joined along the entire length of their wall; outer cells with straight or slightly convex free margins along their entire length; inner cells with conical-rounded poles, sometimes with accessory spines; cenobia 4-8 or more celled .... *D.   maximus*
36. With 2 diagonally arranged main spines .... *D.   armatus* var. *bicaudatus*
37. Cylindrical or oval-cylindrical cells, joined along most of their lateral margins; rounded or broadly rounded poles, longitudinal ribs from pole to pole or only at the apexes of the cells20
   38. Short accessory spines absent ....*D.   armatus* var. *armatus*
   39. Short accessory spines present21
40. 1-3 short accessory spines on the poles.....................................*D.   armatus* var. *spinosus*
41. Approximately oval-fusiform cells, partially joined by
   their lateral margins; .... pronounced poles22
   42. Cell poles pronounced, rounded, without ribs or denticles; main spines located in the middle part of the poles.. *D.   protuberans*
   43. Poles pronounced, truncated, sometimes ribs and denticles; main spines located at the angle of the cell poles ....(D. *opoliensis*) 23
44. Accessory polar denticles present, short ribs present..............*D.   opoliensis* var. *carinatus*
45. .... Absent polar denticles and accessory ribs24
   46. Markedly alternating cells .... *D.   opoliensis* var. *opoliensis*
   47. Cells more or less in a line... *D.   opoliensis* var. *mononensis*
48. Margins of outer cells with 2 spines located in their middle part (equatorially)   *D. flavescens* var. *breviaculeatus*
49. Lateral margins of outer cells with 1 to numerous spines located in ... longitudinal series26
   50. 1(-2) lateral spines on outer cells27
   51. 2 or more lateral spines on the outer cells29
52. Internal cell poles with long spines...*D.pleiomorphus*
53. Internal cell poles with short spines28
   54. 1-2 spines of the same size on the outer margin .... *D. spinosus*
   55. 1-2 spines of unequal size on the outer margin...................................... *D. abundans*
56. (2-)3(-5) lateral spines on outer cells; polar spines of equal length ....*D.   subspicatus*
57. 3-8(-20) lateral spines on outer cells; polar spines of unequal length and diagonally symmetrical *D. gutwinskii*

Desmodesmus abundans **(Kirchner) Chodat** (Fig. 33-38)
Algological Studies 96: 1. 2000.
Basionimo: *Scenedesmus caudatus* **Corda f.** *abundans* **Kirchner** *in* Cohn, Kryptogamen-
   Floravon Schlesien2(1): 98. 1878.
Synonyms: *Scenedesmus abundans* (Kirchner) Chodat
   Scenedesmus quadricauda (Turpin) Brébisson var. *quadrispina* (Chodat) G.M.Smith
   Scenedesmus quadricauda (Turpin) Brébisson var. *parvus* G.M. Smith
   Flat cenobium formed by 2-4 linearly arranged cells; elliptical to oblong cells, rounded poles, outer

cells with 1 spine on each pole and 1-2 spines on the margin, 1 shorter and 1 longer spine, outer cells may vary showing 1 convexity on the margin, inner cells with small spines that may be present or absent; cells can be 2 times longer than wide, equal in length to the polar spines, cell length 4.4-20.3 pm, width 2.9-9.0 pm, length of polar spines 4-8.5 pm, length of marginal spines 2-6 pm; cells can also have small polar spines, proportional to the width of the cell; parietal chloroplastid, 1 pyrenoid.

Geographical distribution in the State of Sao Paulo

IN LITERATURE: **Municipios of Itirapina, Piratininga** and **Sao Paulo** [SANT'ANNA 1984: 246, fig. 159, as S. *quadricauda* (Turpin) Brébisson var. *quadrispina* (Chodat) G.M. Smith]. **Municipio de Sao José dos Campos** [CARDOSO 1979: 98, fig. 126-127, as S. *quadricauda* (Turpin) Brébisson var. *quadrispina* (Chodat) G.M. Smith]. **Municipality of Sao Paulo** [SANT'ANNA 1984: 245, fig. 160, as S. *quadricauda* (Turpin) Brébisson var. *parvus* G.M. Smith].

EXAMINED MATERIAL: **Municipality of Barretos**, lake region, with grasses and Cyperaceae, 28-II-1990, *L.H.Z. Branco* (SP255772). **Municipality of Pirassununga**, SP-225, km 23, Bairro Cascalho, lake, ?-XII-1973, *P.A.C. Senna* (SP123900). **Municipality of Presidente Venceslau**, SP-563, km ?, marsh with aquatic vegetation, clear water, periphyton, 21-VII-1991, *M.C. Bittencourt-Oliveira* (SP255757). **Municipality of Rio Claro**, SP-310, km 156, flooded, 10-V-1973, *C.E.M. Bicudo & P.A.C. Senna* (SP104728). **Municipio de Sao Paulo**, Santo Amaro, lagoon near Av. Washington Luiz, nº 3669, 01- VI-1967, *B. Skvortzov* (SP104098); Parque Estadual das Fontes do Ipiranga, Lago do IAG, 03-VIII-1998, *I.S. Vercellino*, 23° 38'08"S e 23° 40'18"S, 46°36'48'W e 46°38'00'W (SP399779). **Municipality of Tupa**, no precise location, 20-VII-1973, *D.M. Vital* (SP130789).

Comment

Morphologically, *Desmodesmus abundans* (Kirchner) Hegewald is easily confused with *Desmodesmus spinosus* (R. Chodat) Hegewald, however, the difference between the two species lies in the fact that *Desmodesmus abundans* (Kirchner) Hegewald has two *spines of* unequal size on the outer margins of the cells. The spines can also occur singly or, more rarely, in threes. In *Desmodesmus spinosus* (R. Chodat) Hegewald, the marginal *spines* are practically the same size, are much larger in number and are distributed throughout the cell wall of the entire cenobium.

According to HEGEWALD (1978) in *Scenedesmus* and HEGEWALD ETAL. (2001) in *Desmodesmus,* two subsections are recognized within the *Desmodesmus* Section, namely: Subsec. *Subspicati* and Subsec. *Spinosi.* In the latter subsection, the marginal cells have spines on their lateral parts and the internal cells can also have short or long spines. Costelloid structures have so far only been observed in *D. spinosus* (Chodat) Hegewald.

Based on KOMÁREK & LUDVÍK (1972), the three most frequent species of this *Spinosus* subsection [*D. abundans* (Kirchner) Hegewald, *D. spinosus* (Chodat) Hegewald and *D. subspicatus (*Chodat) Hegewald & Schmidt] are easily distinguished using the scanning electron microscope. HEGEWALD *ET AL.* (2001) added the characteristics of electron microscopy to the diagnostic ones observed under an optical microscope. According to the authors, two groups of species can first be differentiated, namely: (7) cells with equatorial lateral spines ("tenuispina"); but if there are only two equatorial spines on the marginal cells, it is *D. flavescens (Chodat). flavescens* (Chodat) Hegewald; and if there are equatorial spines on both the marginal and inner cells, it is *D. asymmetricus* (Schröder) Hegewald; and (2) cells with longitudinally arranged lateral spines ("abundans"). Always taking into account the variability of populations, we find species with 1-2 (usually 1) lateral spines on the marginal cells (D. *abundans* (Kirchner) Hegewald. *D. spinosus* (Chodat) Hegewald and *D. pleiomorphus* (Hindák) Hegewald) or even with five (usually 2-3) lateral spines on the marginal cells [D. *subspicatus* (Chodat) Hegewald & Schmidt and *D. kissii* (Hortobágyi) Hegewald].

Desmodesmus abundans (Kirchner) Hegewald differs easily from *D. pleiomorphus* (Hindák) Hegewald because in the latter the inner cells have large spines, whereas in *D. abundans* (Kirchner) Hegewald there are only very short spines. It is therefore difficult to distinguish *D. abundans* (Kirchner) Hegewald from *D. spinosus* (Chodat) Hegewald. In this case, the representatives of the latter species have lateral ribs and there is also a tendency for the cenobia to break up into solitary cells.

Desmodesmus subspicatus (Chodat) Hegewald & Schmidt has oblong, somewhat elongated cells with truncated poles and *D. kissii* (Hortobágyi) Hegewald has oval-elliptical cells. Other species with numerous lateral spines are *D. multiformis* (Hegewald & Hindák) Hegewald and *D. multivariabilis* Hegewald. Schmidt. Braband & Tsarenko. However, both are known so far only from material in cultivation (HEGEWALD *ET al.* 2005).

In all the species described so far with these characteristics. the lateral spines are very stable. both in number and size. so that. under an optical microscope. recognizing such species is really difficult. giving rise to different interpretations.

Representatives of *D. abundans* (Kirchner) Hegewald were found in seven localities in the state of São Paulo. *Desmodesmus abundans* (Kirchner) Hegewald is not so easy to identify. We found specimens with cenobiums of two or four cells, with a long thorn on the poles of the outer cells and at least one thorn in the center of the margin of the outer cell and small thorns on the inner cells. Morphological variation was detected with regard to the shape of the cell, which sometimes appeared oblong, sometimes elliptical with a very pronounced convexity in the center of the outer margin, where the thorn on the outer margin of the cell comes from. Morphological variation was also detected with regard to the thickness of the spines, which sometimes appear very thin and sometimes quite thick. Care should be taken when making taxonomic identifications, as *D. abundans* (Kirchner) Hegewald is very similar to *D. spinosus* (Chodat) Hegewald and can sometimes be easily confused with each other.

Scenedesmus aculeolatus Reinsch (Fig. 141-142)

Journal of the Linnean Society of London, Botánica series 16: 238, pl. 6, fig. 1-2. 1877.

Cenobium flat, formed by 2-4 cells, arranged linearly; cells elliptical, poles rounded, cells in contact for almost the entire longitudinal length, outer cells and inner cells with 1-3 small spines at the poles, may show thickening of the cell wall; cell length 5.41-15.25 pm, diameter 1.9-5.1 pm, length of polar spines 1-2 pm; parietal chloroplastidium, 1 pyrenoid.

Geographical distribution in the state of Sao Paulo

IN THE LITERATURE: nothing. First mention of the species in the state of Sao Paulo.

EXAMINED MATERIAL: **Municipality of Bragança Paulista**, 3 km northeast of the city of Bragança Paulista, pond, 20-VI-1973, *D.M. Vital* (SP113524). **Municipality of Campos de Jordao,** Alagoinha neighborhood, lake, 11-III-1973, *M.M. Sakane* (SP130445). **Municipality of Cananéia**, Ilha Comprida, 120 m from the sea, lagoon, 07-III-1975, *M. Vital* (SP130813). **Municipality of Divinolândia**, DVL-040, 13 km from the junction with the SP-344, waterfall, 08-VIII-2000, *C.E.M. Bicudo, L.A. Carneiro & S.M.M. Faustino* (SP365697). **Municipality of Pindamonhangaba,** BR-116, km 270-271, lagoon, 21-V-1966, *C.E.M. Bicudo* (SP96949); Sao Joao farm, Sao Joao lagoon, 21-V-1966, *C.E.M. Bicudo* (SP96950). **Municipality of Piracaia**, SP-36, km 101-102, Jacareí river reservoir, 24-IV-2000, *C.E.M. Bicudo & C.I. Santos* (SP365699). **Municipality of Santo André,** Estação Biológica do Alto da Serra de Paranapiacaba, lake near the headquarters, 20-I-1966, *C.E.M. Bicudo* (SP96909). **Municipality of Tambaú**, Clube de Tambaú, dam, 23-VI-1973, *D.M. Vital* (SP113574).

Comment

Scenedesmus aculeolatus Reinsch can be confused with *Desmodesmus arthrodesmiformis* (Schroder) An *et al.* as both species can show thickening of the cell poles, as well as differences in the arrangement of the polar spines, the number of spines and the shape of the cells.

The species is characterized, according to its original description in REINSCH (1877), by its oblong-cylindrical cells arranged in aligned cenobia and by the obtuse-rounded poles armed with three to five denticles (HEGEWALD & SILVA 1988: 48, fig. 58). *Scenedesmus aculeolatus* Reinsch appears frequently but, according to KOMAREK & FOTT (1983), it is generally identified with *Desmodesmus denticulatus* (Lagerheim) Hegewald var. *linearis* (Hansgirg) Hegewald (= *Scenedesmus denticulatus* Lagerheim var. *linearis* Hansgirg).

ECHENIQUE *ET al.* (2004) described the shape of the cells as oval-cylindrical and arranged in a line, with short spines or denticles at the cell ends. These authors mentioned that they had not observed ribs in the specimens they examined. In fact, HEGEWALD & SILVA (1988) mentioned the absence of ribs in the original description of *S. aculeolatus* Reinsch.

HEGEWALD (2000) did not make the new combination for *Scenedesmus aculeolatus* Reinsch, so this species is still within the genus *Scenedesmus,* however, the species needs a new combination to be within the genus *Desmodesmus* due to the fact that the species has spines and ornamentation on the cell wall.

Desmodesmus armatus (R. Chodat) Hegewald var. *armatus* (Fig. 42-46)

Algological Studies 96: 2. 2000.

Basionimo: *Scenedesmus hystrix* Lagerheim var. *armatus* R. Chodat, Algues Vertes de Suisse. 215, fig. 140. 1920.

Synonyms: *Scenedesmus armatus* (Chodat) Hegewald var. *armatus*

        Scenedesmus pseudoarmatus Hortobágyi

        Scenedesmus ellipsoideus Chodat

Cenobium flat, formed by 2-4 cells, cells linearly arranged, elliptic to oblong, poles rounded to acute, outer cells with 1 spine on each pole, inner cells with small spines, cell wall with continuous or interrupted median longitudinal ribs and rosettes (rarely); cell length 5.4-15.9 pm (literature 14.0-18.0 pm), width 1.9-5.3 pm, (literature 5.0-8.0 pm), polar spines 4-14 pm long (literature 10.0-20.0 pm); parietal chloroplastid, 1 pyrenoid.

Geographical distribution in the state of Sao Paulo

IN LITERATURE: **Municipality of Sao Paulo** (FERRAGUTETAL. 2005: 153, fig. 75). **Municipality of Luiz Antonio** [PERES & SENNA 2000: 474, fig. 32, as *Scenedesmus armatus* (Chodat) Hegewald var. *boglariensis* Hortobágyi]. **Municipality of Ribeirao Preto** (SILVA 1999: 291, fig. 64, as *Scenedesmus longispina* R. Chodat). **Municipality of São Paulo** (SANT'ANNAETAL. 1989: 96, fig. 83, as *Scenedesmus ellipsoideus* Chodat].

EXAMINED MATERIAL: **Municipality of Arujá**, Clube Fiscal do Brasil, lake, 01- V-1975, *L. Sormus* (SP130815). **Municipality of Barretos, lake** region, with grasses and Cyperaceae, 28-II-1990, *L.H.Z. Branco* (SP255772). **Municipality of Guaratinguetá**, Clube dos 500, lake, 01-IV-1966, *C.E.M. Bicudo* (SP96965). **Municipality of Juquiá**, BR-116, km 165, lake, 01-III-1973, *C.EM. Bicudo & L. Sormus* (SP113664). **Municipality of Mairipora**, Santa Inés road, Mairipora reservoir, 12-IV-1992, *M.C. Bittencourt-Oliveira* (SP239242). **Municipality of Moji das Cruzes**, SP-88, 1 km before Moji das Cruzes, direction Salesópolis- Moji das Cruzes, 21-II-1989, *A.A.J. de Castro & C.E.M. Bicudo* (SP188211). **Municipality of Pindamonhangaba**, BR-116, km 270-271, lagoon, 21-V-1966, *C.E.M. Bicudo* (SP96949); Fazenda Sao Joao, lagoon Sao Joao, 21-V-1966, *C.E.M. Bicudo* (SP96950). **Municipio de Presidente Venceslau**, SP-563, km ?, marsh with aquatic vegetation, clear water, 21-VII- 1991, col. *M.C. Bittencourt-Oliveira*, periphyton (SP255757). **Municipio de Santo André**, Estagao Biológica do Alto da Serra de Paranapiacaba, lake near the headquarters, 20-I-1966, *C.E.M. Bicudo* (SP96909). **Municipio de Sao Paulo**, Horto Florestal, lake, 16-VII-1962, *C.E.M. Bicudo & R.M.T. Bicudo* (SP96850); Parque Estadual das Fontes do Ipiranga, Lago do IAG, *I.S. Vercellino*, 03-VIII-1998, 23°38'08"S and 23°40'18"S, 46°36'48"W and 46°38'00"W (SP399779); 06-III-1999, I.S. *Vercellino*, 23°38'08"S and 23°40'18"S, 46°36'48'W and 46°38'00"W (SP399780); Lago das Gargas, M. *Borduqui*, 12-VII-2006, 23°38'08"S and 23°40'18"S, 46°36'48"W and 46°38'00"W (SP399781); 17-I-2007, 23°38'08"S and 23°40'18"S, 46°36'48"W and 46°38'00"W (SP399782). **Municipality of Tambaú**, Clube de Tambad, dam, 23-VI-1973, *D.M. Vital* (SP113574). **Municipality of Tremembé**, SP-442, 13.2 km before Taubaté, lake on the right, direction Pindamonhangaba-Taubaté, with *Utricularia, Typha* Cyperaceae, periphyton, 24-IV-1990, *A.A.J. de Castro & C.E.M. Bicudo* (SP255731). **Municipio de Uchoa**, SP-310, km 410.7, 30 km before Sao José do Rio Preto, pond with Poaceae, clay bottom, periphyton, 10-IV-1990, *D.C. Bicudo & C.E.M.Bicudo* (SP255729).

Comment

Desmodesmus armatus (Chodat) Hegewald was not easy to identify. This species is morphologically very similar to *Desmodesmus communis* (Hegewald) Hegewald, from which it differs in the presence of ribs. *Desmodesmus armatus* (Chodat) Hegewald shows great morphological variation both in the size of the cells and in the length of the spines and the presence of ribs. The polar spines of the latter species are generally quite long and thick, however, they can vary and there can also be small polar spines on the inner cells. The ribs may also be present, and may appear whole, cutting across the entire longitudinal axis, or may be fragmented.

Desmodesmus armatus (Chodat) Hegewald together with *D. opoliensis* (P. Richter) Hegewald constitute, according to the most recent criteria, species in a broad sense, as they include a large list of heterotypic synonyms (HINDÁK 1990, HEGEWALD 2000). *Desmodesmus armatus* (Chodat) Hegewald is fundamentally characterized by the presence of longitudinal ribs on the cell faces as well as spines and decorations formed by the outermost layer of the cell wall. The taxonomic varieties of *Desmodesmus opoliensis* (P. Richter) Hegewald can also have ribs on the sides [D. *opoliensis* (P. Richter) Hegewald var. *carinatus* (Lemmermann) Hegewald]. The ultrastructure of the cell wall is also different (KOMÁREK & LUDVÍK, 1972), although they can be confused with the juvenile stages of *D. armatus* (Chodat) Hegewald. However, under optical microscopy they are easily separated since the cells of *D. opoliensis* (P. Richter) Hegewald have truncated marginal cell poles and in *D. armatus* (Chodat) Hegewald they are rounded.

Desmodesmus armatus (Chodat) Hegewald is, according to HINDÁK (1990) and HEGEWALD (2000), one of the species whose diagnostic characters show wide variability. HEGEWALD (2000) considered and recognized seven taxonomic varieties of D. *armatus* (Chodat) Hegewald, three of which are represented in the state of Sao Paulo.

PERES & SENNA (2000) identified S. *armatus* (Chodat) Hegewald var. *boglariensis* Hortobágyi in a paper on the Chlorophyta of Lagoa do Diogo, but on examining the illustrations in this paper we concluded that it was D. *armatus* (Chodat) Hegewald var. *armatus*.

SILVA (1999) identified *Scenedesmus longispina* R. Chodat from the phytoplankton of a eutrophic reservoir in Ribeirao Preto; however, an examination of the illustrations in this study showed that it was *D. armatus* (Chodat) Hegewald var. *armatus*.

Representatives of *D. armatus* (Chodat) Hegewald were collected in 14 municipalities in the state of Sao Paulo. *Desmodesmus armatus* (Chodat) Hegewald was easily found in the samples examined, but not very easy to identify. The species showed relatively large morphological variation with regard to the size of the cells, the size of the spines and the number of spines on the inner cells. However, one character that was always stable in the populations studied was the presence of ribs, although this character also varied in the species. It was noted that extreme care is needed when identifying this species because, as well as showing considerable morphological variation, it is often quite difficult to differentiate its representatives from those of *D. communis* (Hegewald) Hegewald.

Desmodesmus armatus (Chodat) Hegewald var. *bicaudatus* (Guglielmetti) Hegewald (Fig. 47-50) Algological Studies 96: 4. 2000.

Basionimo: *Scenedesmus acutiformis* Schroder var. *bicaudatus* Guglielmetti, Nuova Notarisia21: 31. 1910.

Synonyms: *Scenedesmus semipulcher* Hortobágyi
    Scenedesmus armatus (Chodat) Hegewald var. *bicaudatus* (Guglielmetti) Chodat)
    Scenedesmus decorus Hortobágyi var. *bicaudato-granulatus* (Hortobágyi) Uherkovich

Flat cenobium formed by 2-4 cells; cells arranged linearly, elliptical, poles rounded to acute, outer cells with 1 spine only on 1 of the poles, arranged in opposite directions (diagonally), tiny spines along the entire length of the margin of the outer cells (like a comb), inner cells with or without small short spines, longitudinal ribs whole or fragmented, median along the entire length of the cell, ribs may also appear as 1 longitudinal stria; cell length 6.0-13.9 pm, width 1.5-6.0 pm, polar spines 5.0-13.0 pm long; parietal chloroplastid, 1 pyrenoid.

Geographical distribution in the State of Sao Paulo

IN LITERATURE: **Municipality of Arujá** [SANT'ANNA 1984: 215, fig. 139-140, as *Scenedesmus decorus* Hortobágyi var. *bicaudato-granulatus* (Hortobágyi) Uherkovich]. **Municipality of Luiz Antonio** (PERES & SENNA 2000: 473, fig. 31). **Municipality of Sao Paulo** [MOURA 1996: 49, fig. 18, as *Scenedesmus semipulcher* Hortobágyi; FERRAGUT *ET AL.* 2005: 153, fig. 76; TUCCI *ET AL.* 2006: 165, fig. 51, as *Scenedesmus semipulcher* Hortobágyi].

EXAMINED MATERIAL: **Municipality of Arujá**, Clube Fiscal do Brasil, lake, 01- V-1975, *L. Sormus* (SP130815). **Municipality of Guaratinguetá**, Clube dos 500, lake, 01-IV- 1966, *C.E.M. Bicudo* (SP96965). **Municipality of Rio Claro**, SP-310, km 156, flooded, 10-V- 1973, *C.E.M. Bicudo & P.A.C. Senna* (SP104728). **Municipio de Itaporanga**, SP-255, km 358, river, 26-VII-2000, *S.M.M. Faustino, & S.P. Schetty*, 23°42'24,3"S, 49°28'15,6"W, conductivity 20 pS cm^{-1}, pH 7,1 (SP355358). **Municipality of Paraguayo Paulista**, SP-284, km 457, stream after dam, no vegetation on banks or aquatic, 20-VII-1991, *M.C. Bittencourt-Oliveira* (SP239085). **Municipality of Pindamonhangaba**, BR-116, km 270-271, pond, 21-V-1966, *C.E.M. Bicudo* (SP96949). **Municipio de Sao Paulo**, Parque Estadual das Fontes do Ipiranga, Lago das Garcas, *M. Borduqui*, 12-VII-2006, 23°38'08"S and 23°40'18"S, 46°36'48"W and 46°38'00"W (SP399781); 17-I-2007,M. *Borduqui*, 23°38'08"S and 23°40'18"S, 46°36'48"W and 46°38'00"W (SP399782).

Comment

Desmodesmus armatus (Chodat) Hegewald var. *bicaudatus* (Guglielmetti) Hegewald differs from the typical variety of the species by its bicaudate character. KOMÁREK & FOTT (1983) included the species in *Scenedesmus semipulcher* Hortobágyi. Recently, the new combination *Desmodesmus armatus* var. *bicaudatus* (Guglielmetti) Hegewald was made in HEGEWALD (2000). We agree with TUCCI *ET al.* (2006) when they stated that the outer cells of the cenobium of *Desmodesmus armatus* (Chodat) Hegewald var. *bicaudatus* (Guglielmetti) Hegewald have a long spine at one of the poles and the inner cells have a longitudinal groove. This rib is in the form of a groove that runs along all or almost all of the length of the cell and is a distinctive feature of the species. In addition to these characteristics, the specimens from the state of São Paulo have tiny spines that occur very close together, running the entire length of the margin of the outer cells of the cenobium, resembling a comb, a feature that is also characteristic of *Desmodesmus armatus* (Chodat) Hegewald var. *bicaudatus* (Guglielmetti) Hegewald.

Desmodesmus armatus (Chodat) Hegewald var. *bicaudatus* (Guglielmetti) Hegewald is based on *Scenedesmus acutiformis* Schröder var. *bicaudatus* Guglielmeti and is unique in that it has only two spines located diagonally opposite each other on each outer cell of the cenobium. The 'status' *bicaudatus* occurs frequently in other species of the genus. However, HEGEWALD (2000) never considered this feature to be a good character for separating taxonomic varieties and, as such, it was nominally transferred. Among the most notable synonyms of this species are: *Scenedesmus bicaudatus* (Chodat) Chodat, *.S' cristatus* Uherkovich and S. *semipulcher* Hortobágyi. *Scenedesmus decorus* Hortobágyi var. *bicaudatus-granulatus* (Hortobágyi) Uherkovich identified by SANT'ANNA (1984) was included by HEGEWALD (2000) in *Desmodesmus armatus (*Chodat) Hegewald var. *bicaudatus (*Guglielmetti) Hegewald.

Representatives of *Desmodesmus armatus* (Chodat) Hegewald var. *bicaudatus* (Guglielmetti) Hegewald were found in seven localities in the state of São Paulo. The species was considered common in the state's waters and easy to identify taxonomically. However, care should be taken not to confuse it with *Desmodesmus intermedias* (R. Chodat) Hegewald var. *acutispinus* (Roll) Hegewald, which differs from *Desmodesmus armatus* (Chodat) Hegewald var. *bicaudatus* (Guglielmetti) Hegewald in that it does not have comb-like denticles in the outer cells of the cenobium. Specimens were found with cenobiums made up of two or four cells. The shape of the cell was also considered to be a very variable character within the species, with the size and shape of the outer cells varying, which in some specimens were elliptical, but with the poles approximately truncated. The size of the spines on the poles of the outer cells also varied, as did the presence or absence of tiny spines on the poles of the inner cells. The presence or absence of ribs can also be a variable character within the species. However, the presence of tiny spines located very close together, which run along the entire length of the margin of the outer cells like a comb, was the most stable character of this variety in the populations examined from the state of Sao Paulo.

***Desmodesmus armatus* (Chodat) Hegewald var. *spinosus (*Fritsch & Rich) Hegewald** (Fig. 51-52)
Algological Studies 96: 5. 2000.
Basionimo: *Scenedesmus armatus* (Chodat) Hegewald *var. spinosus* Fritsch & Rich,
 Transactions of the Royal Society of South Africa 18: 31, fig. 5a-c. 1929.
Synonyms: *Scenedesmus praetervisus* Chodat
  Scenedesmus denticulatus Lagerheim var. *linearis* Hansgirg f. *costato-granulatus*
   (Hortobágyi) Uherkovich
  Scenedesmus denticulatus Lagerheim var. *linearis* Hansgirg f. *granulatus* Hortobágyi
Flat cenobium formed by 4-8 cells, cells arranged linearly, ovate to oblong, poles rounded to acute, outer and inner cells with 1-3 spines at the poles, longitudinal median ribs along the entire length of the cell wall, may or may not have tiny spines located very close to each other forming a comb, cells ca. 3 times longer than wide, cell length 10.2-20.3 pm, width 3.0-7.8 pm, polar spines 2-3 pm long; parietal chloroplastid, 1 pyrenoid.

Geographical distribution in the State of Sao Paulo
IN LITERATURE: **Municipality of Mogi das Cruzes, Porangaba, Rio Claro, Sao Carlos, Sorocaba** and **Sumaré** [SANT'ANNA 1984: 221, fig. 144, as *Scenedesmus denticulatus* Lagerheim var. *linearis* Hansgirg f. *costato-granulatus* (Hortobágyi) Uherkovich]. **Municipios de Itirapina, Jaú** e **Sumaré** [SANT'ANNA 1984: 223, fig. 143, as *Scenedesmus denticulatus* Lagerheim var. *linearis* Hansgirg f. *granulatus* Hortobágyi].

EXAMINED MATERIAL: **Municipality of Itu**, SP-280, km 77, lake, 11-V-1977, *C.R. Leite* (SP139733). **Municipality of Lengóis Paulista**, SP-300, km 299.5, lake at the entrance to the town, Lengois river, 22-II-1992, *C.M. Bicudo & D.C. Bicudo* (SP239236). **Municipality of Rio Claro**, Horto Florestal, lake, 31-I-1975, *O.A. da Silva* (SP123867). **Municipality of Sao Paulo,** Parque Estadual das Fontes do Ipiranga, Lago das Ninféias, col. *T.R. Santos,* 03-VIII- 2007, 23°38'08"S e 23°40'18"S, 46°36'48"W e 46°38'00"W (SP399783). **Municipio de Ubatuba**, no precise location, 29-I-1966, *O. Montes &R.R. Martins* (SP96890).

Comment
Desmodesmus armatus (Chodat) Hegewald var. *spinosus* Fritsch & Rich's main characteristics, apart from the ribs on the cell faces, are the short, vigorous spines, both on the poles of the marginal cells and on those of the inner cells. The spines on the marginal cells sometimes have diagonal symmetry. Next to the spines, there may be warts or granules, although this is optional.

According to HEGEWALD (2000), this variety includes *Scenedesmus praetervisus* Chodat and S. *soli* Hortobágyi. *Scenedesmus denticulatus* Lagerheim var. *linearis* Hansgirg f. *costato-granulatus* (Hortobágyi) Uherkovich and *Scenedesmus denticulatus* Lagerheim var. *linearis* Hansgirg f. *granulatus* Hortobágyi were included by HEGEWALD (2000) in *Desmodesmus armatus (*Chodat) Hegewald var. *spinosus* (Fritsch & Rich) Hegewald.

Representatives of *Desmodesmus armatus* (Chodat) Hegewald var. *spinosus* (Fritsch & Rich) Hegewald were found in five locations in the state of Sao Paulo, but it was not a very common species due to the small number of individuals found in each sample analyzed.

The presence of spines on the poles of the inner cells of the cenobium is a very variable feature within the variety. In addition to this, the tiny spines located very close together and running along the entire length of the margin of the outer cells like a comb was also a highly variable feature of this variety. The most stable feature of the species was the presence of ribs.

Desmodesmus arthrodesmiformis (Schröder) An, Friedl & Hegewald (Fig. 53-55)

In Hegewald, Algological Studies 96: 7. 2000.

Basiönimo: *Scenedesmus arthrodesmiformis* Schröder, Berichte der Deutschen Botanisches Geselcgaft 38: 134, fig. 6. 1920.

Synonym: *Scenedesmuspulloideus* Hegewald

Cenobium flat, formed by 4 or 8 cells; cells linearly arranged, elliptic, elliptic-fusiform or oval-cylindrical, poles rounded, gently converging at the poles, truncated or semi-rectangular; outer and inner cells with 1 small spine which may also be missing (rare); spines facing downwards (towards each other in the same cell), cell length 6.2-17.2 pm, width 1.8-5.0 pm, polar spines 2-3 pm long; parietal chloroplastid, 1 pyrenoid.

Geographical distribution in the state of Sao Paulo

IN LITERATURE: nothing. First mention of occurrence in the state of Sao Paulo.

EXAMINED MATERIAL: **Municipio de Barretos**, Barretos, in the city, lake region, with grasses and Cyperaceae, 28-II-1990, *L.H.Z. Branco* (SP255772). **Municipality of Brodowski**, highway, km 7, on the left, direction Brodowski-Jardinópolis, in front of km 7, marsh with Cyperaceae and *Typha,* 16-XI-1991, *A.A.J. de Castro* (SP239098). **Municipio de Caconde**, Represa de Caconde, SP-344, km 291, col. *C.E.M. Bicudo, L.A. Carneiro & S.M.M. Faustino,* 08-VIII-2000, 21°34'39.9" S, 46°37'31.0" W, conductivity 30 pS cm⁻¹ , pH 8.0 (SP355356). **Municipality of Guaratinguetá**, Clube dos 500, lake, 01-IV-1966, *C.E.M. Bicudo* (SP96965). **Municipality of Miguelópolis**, dam, 30-V-2000, *C.E.M. Bicudo & D.C. Bicudo* (SP365690). **Municipality of Pindamonhangaba** BR116, km 270-271, lagoon, 21-V-1966, *C.E.M. Bicudo,* (SP96949); Säo Joäo farm, Sao Joäo lagoon, 21-V-1966, *C.E.M. Bicudo* (SP96950). **Municipality of Pirassununga**, SP-225, km 23, Cascalho neighborhood, lake, ?- XII-1973, *P.A.C. Senna* (SP123900). **Municipio de Pitangueiras**, acude, SP-322, km 368, 16-VII-2000, C.E.M. Bicudo, S.M.M. Faustino, & L.L. Morandi, 20°59'30,5" S, 48°14'01,1" W, conductivity 40 pS cm⁻¹ , pH 6,5 (SP355382). **Municipality of Rio Claro**, SP-310, km 156, flooded, 10-V-1973, *C.E.M. Bicudo & P.A.C. Senna* (SP104728); Horto Florestal, lake, 31-I-1975, *O.A. da Silva* (SP123867). **Municipality of Sao Bernardo do Campo,** Grande creek, Billings reservoir, 05-X-1972, *C.R. Leite* (SP130432).

Comment

Certain morphological expressions of *Desmodesmus arthrodesmiformis* (Schröder) An *et al.* can easily be confused with those of *Desmodesmus denticulatus* (Lagerheim) An *et al.* var. *linearis* (Hansgirg) Hegewald [formerly *Scenedesmus brevispina* (G.M. Smith) Chodat]. These two species differ in their cell dimensions, as well as the orientation of the spines, which in *Desmodesmus arthrodesmiformis* (Schröder) An *et al.* are projected downwards (convex) and rectilinear, parallel to the longitudinal axis of the cell in *Desmodesmus denticulatus (*Lagerheim) An *et al.* var. *linearis (*Hansgirg) Hegewald. *Desmodesmus arthrodesmiformis* (Schröder) An *et al.* shows considerable polymorphism, including elliptical, elliptical-fusiform or oval-cylindrical cells, whose poles can be rounded to slightly convergent, truncated or semi-rectangular.

The species identified in HORTOBÁGYI (1967) as *Scenedesmus arthrodesmiformis* Schröder, now *D. arthrodesmiformis* (Schröder) An *et al.* has a cell shape and certain cell structures that resemble denticles those of *D. denticulatus* (Lagerheim) An *et al.* var. *sphenisciformis* Hortobágyi. The same author also mentioned that the latter variety may have ribs, as well as granulations in the cell walls. According to HEGEWALD *ET AL.* (1980), S. *pulloideus* Hegewald, now synonymous with D. *arthrodesmiformis* (Schröder) An *et al. is* very reminiscent of *Desmodesmus armatus* (Chodat) Hegewald in terms of cell wall ornamentation, the difference between the two lying in the cenobia with small spines and the production of mucilage in cultivation in D. *arthrodesmiformis* (Schröder) An *et al.*

Representatives of D. *arthrodesmiformis* (Schröder) An *et al.* were found in 10 localities in the state of São Paulo, but its taxonomic identification is not so simple because it is very similar to D. *denticulatus* (Lagerheim) An *et al.* var. *linearis* (Hansgirg) Hegewald, from which they differ in cell shape. *Desmodesmus arthrodesmiformis* (Schröder) An *et al.* is a very morphologically variable species with regard to the shape of its cells and also the number of spines at the cell poles. In addition, it can show thickening of the cell poles (rare) and a cell wall with granulation along the entire length of the cell.

Desmodesmus brasiliensis (Bohlin) Hegewald (Fig. 56-59)

Algological Studies 96: 7. 2000.

Basiönimo: *Scenedesmus brasiliensis* **Bohlin**, Bihang till K. Svenska vetenskapsakademiens handlingar: sér. 3,23(7): 22, pl. 1, fig. 33a-e. 1897.

Synonym: *Scenedesmus brasiliensis* Bohlin var. *norvegicus* Printz

Flat cenobiums formed of 2 or 4 cells, cells arranged linearly, elliptical to oblong, poles rounded; outer and inner cells with 1-2(-3) small spines which may or may not appear on all cells of the cenobium; cell wall smooth or granular throughout, median longitudinal ribs along the entire length of the cell or fragmented; outer cells with or without rosettes; outer cells usually with tiny spines at the margin (comb-like), with a membrane surrounding the comb; cell length 7.0-21.3 pm, width 2.0-5.9 pm, polar spines 1-3 pm long; parietal chloroplastid, 1 pyrenoid.

Geographical distribution in the State of Sao Paulo

IN LITERATURE: **Municipality of Sao Paulo** [HOEHNE 1948: 16, fig. 29, as *Scenedesmus brasiliensis* Bohlin], **Municipality of Luiz Antonio** [PERES & SENNA 2000: 474, fig. 30, as *Scenedesmus hystrix* Lagerheim var. hystrix]. **Municipio de Sao Carlos** [SCHWARZBOLD 1992: 112, fig. 2, as *Scenedesmus denticulatus* Lagerheim].

EXAMINED MATERIAL: **Municipality of Campos de Jordao**, Alagoinha neighborhood, lake, 11-III-1973, *M.M. Sakane* (SP130445). **Municipality of Itu**, SP-280, km 77, lake, 11-V-1977, *C.R. Leite* (SP139733). **Municipality of Juquiá**, BR-116, km 160, flooded, 01-III-1973, *C.E.M. Bicudo &L. Sormus* (SP113672). **Municipality of Lengóis Paulista**, SP-300, km 299.5, lake at the entrance to the town, Lengois river, 22-II-1992, *C.E.M. Bicudo & D.C. Bicudo* (SP239236). **Municipality of Miracatu**, Jaragatiá, Pettena farm, lake, 01-III-1973, *C.E.M. Bicudo, C.R. Leite & L. Sormus* (SP113673). **Municipality of Moji das Cruzes**, SP-88, km 7475, pond, 18-VI-1973, *C.E.M. Bicudo, C.R. Leite & L. Sormus* (SP113662). **Municipio de Moji Guagu**, no precise location indicated, 17-X-1973, *D.M. Vital* (SP114539). **Municipio de Pedro de Toledo**, rodovia Manoel da Nóbrega, km 370.5, agude, 11-VII-2000, *C.E.M. Bicudo & S.M.M. Faustino* (SP365691). **Municipality of Pindamonhangaba**, BR-116, km 270-271, lagoon, 21-V-1966, *C.E.M. Bicudo* (SP96949); Sao Joao farm, Sao Joao lagoon, 21-V-1966, *C.E.M. Bicudo* (SP96950). **Municipality of Pirassununga**, SP-225, km 23, Cascalho neighborhood, lake, ?-XII-1973, *P.A.C. Senna* (SP123900). **Municipality of Rio Claro**, Horto Florestal, lake, 31-I-1975, *O.A. da Silva* (SP123867). **Municipality of Santo André**, Estagao Biológica do Alto da Serra de Paranapiacaba, stream near the headquarters, 20-I-1966, *C.E.M. Bicudo* (SP96922). **Municipality of Sao Carlos**, SP-310, km 222, Aldeia Conde do Pinhal, stream, 10-V-1973, *C.E.M.Bicudo &L. Sormus* (SP104723). **Municipio de Sao Paulo**, Santo Amaro, lagoon near Av. Washington Luiz n^0 3669, 01-VI-1967, *B. Skvortzov* (SP104098); Parque Estadual das Fontes do Ipiranga, Lago das Ninféias, 03-VIII-2007, *T.R. Santos,* 23°38'08"S e 23°40'18"S, 46°36'48"W e 46°38'00"W (SP399783). **Municipality of Tambaú**, Clube de Tambaú, dam, 23-VI-1973, *D.M. Vital* (SP113574). **Municipality of Ubatuba**, no precise location, 29-I-1966, *O. Montes & R.R. Martins* (SP96890). **Municipio de Urania**, SP-300, 1 km before the town, unspecified location, with Cyperaceae, grasses and *Myriophyllum,* 05-XII-1991, *L.H.Z. Branco* (SP239237).

Comment

Desmodesmus brasiliensis (Bohlin) Hegewald is a very common, cosmopolitan species that shows considerable morphological variation. The elliptical to oblong cells are a stable feature of the species, but the polar spines can vary between one and two, rarely three, in all the cells of the cenobium, both external and internal. This structure is characteristic of the species, however, *Desmodesmus brasiliensis* (Bohlin) Hegewald can easily be confused with *Desmodesmus serratus* (Corda) Bohlin due to the fact that it also has this comb-like structure, however, S. *serratus* (Corda) Bohlin has independent marginal spines, more separated from each other than in D. *brasiliensis* (Bohlin) Hegewald. In addition, the ribs can also vary, either present or absent and appearing whole, cutting across the entire longitudinal axis of the cell, or fragmented. The cell wall can also vary, appearing smooth or with granulations.

Following the concept of HEGEWALD (2000), this is also a species with a relatively broad description, however, characterized by the presence of ribs from slight to very developed on the lateral margins of the cells and denticles (sometimes short spines) only at the cell poles. This description includes 25 taxa as synonyms, such as .S' *alatus* Jao, S *circumfusus* Hortobágyi, *S.pluricostatus* Bourrelly and S. *striatus* Dedusenko. HEGEWALD (2000) recognizes only two varieties in D. *brasiliensis* (Bohlin) Hegewald, namely: var. *brasiliensis* and var. *serrato-perforatus* (Patel & George) Hegewald, the latter characterized by the sub-square shape of its cells, with capitate ends, which leave spaces between them.

SCHWARZBOLD (1992) identified *Desmodesmus denticulatus* (Lagerheim) An *et al.* however, on re-examining the materials he identified, we concluded that it was D. *brasiliensis* (Bohlin) Hegewald.

Representatives of D. *brasiliensis* (Bohlin) Hegewald were recorded in 17 localities in the state of

São Paulo. The species was considered common in the state, as well as being a cosmopolitan species, but not always easy to identify taxonomically, due to the fact that it shows considerable morphological variation in terms of the shape of the cells, the number of polar spines and the size of these spines, as well as the presence or absence of ribs. When present, the ribs may appear whole or fragmented. In addition to this variation, the outer cells may have tiny spines very close to the cell margin, resembling a comb, which may be surrounded by a membrane, but this set is also very variable and may or may not be present. In fact, it can be said that among *Desmodesmus, D. brasiliensis* (Bohlin) Hegewald is considered to be one of the species with the greatest morphological variability. Care should be taken when identifying the varieties and taxonomic forms of D. *brasiliensis* (Bohlin) Hegewald due to the fact that there are morphotypes that are not very distinct from D. *brasiliensis* (Bohlin) Hegewald, which really make it difficult to identify D. *brasiliensis* (Bohlin) Hegewald.

Desmodesmus communis *(*Hegewald) Hegewald** (Fig. 60-63)
Algological Studies 96: 8. 2000.
Basionimo: *Scenedesmus communis* Hegewald, Algological Studies 19: 151, fig. 12-13. 1977.
Synonyms: *Scenedesmus quadricauda* (Turpin) Brébisson *sensu* Chodat
          Scenedesmus quadricauda (Turpin) Brébisson f. *granulatus* Hortobágyi
Cenobium flat, formed of 2-4 cells, cells arranged linearly, elliptical to oblong, poles rounded, outer cells with 1 spine on each pole, inner cells devoid of spines; outer cells with 1 median convexity on outer margin. Cell length 6.1-22.7 pm, width 1.7-8.7 pm, polar spines 2-22.5 pm long; parietal chloroplastid, 1 pyrenoid.

Geographical distribution in the State of Sao Paulo
IN LITERATURE **Municipalities of Arujá, Atibaia, Barra Bonita, Campos do Jordao, Cananéia, Guaratinguetá, Ibirá, Irapua, Itirapina, Itu, Jaú, José Bonifácio, Juquiá, Miracatu, Moji das Cruzes, Pindamonhangaba, Rio Claro, Salesópolis, Santo André, Sao Bernardo do Campo**, Sao **Carlos, Sao Paulo, Sorocaba, Sumaré, Tupa** and **Ubatuba** [SANT'ANNA 1984: 237, fig. 158, as *Scenedesmus quadricauda* (Turpin) Brébisson var. *quadricauda]*. **Municipality of Juquiá** [SANT'ANNA *ET AL.* 1988: 91, fig. 43, as *Scenedesmus quadricauda* (Turpin) Brébisson]. **Municipio de Luiz Antonio** [PERES & SENNA 2000: 474, fig. 33, as *Scenedesmus quadricauda* (Turpin) Brébisson var. *quadricauda]*. *Municipio* **de Municipio de Ribeirao Preto** [SILVA 1999: 291, fig. 66, as *Scenedesmus protuberans* Fritsch & Rich]. **Municipality of Sao Carlos** [SCHWARZBOLD 1992: 112, fig. 1, as *Scenedesmus quadricauda* (Turpin) Brébisson]. **Municipio de Sao José dos Campos** [CARDOSO 1979: 92, fig. 118-123, 125, as *Scenedesmus quadricauda* (Turpin) Brébisson var. *quadricauda]*. **Municipio de Sao Paulo** [XAVIER 1979: 22, fig. 20, as *Scenedesmus opoliensis* P. Richter; SANT'ANNA *ET AL.* 1989: 97, fig. 92-93, as *Scenedesmus quadricauda* (Turpin) Brébisson; GENTIL 2000: 55, fig. 16, as *Scenedesmus quadricauda* (Turpin) Brébisson; FERRAGUT *ETAL.* 2005: 153, fig. 81, as *Desmodesmus opoliensis* (P. Richt) Hegewald var. *opoliensis;* FERRAGUT *ET AL.* 2005: 156, fig. 99, as *Scenedesmus quadricauda* (Turpin) Brébisson *sensu* Chodat]. **Teodoro Sampaio** [BICUDO ET *al.* 1992: 303, fig. 43, as *Scenedesmus quadricauda* (Turpin) Brébisson var. *quadricauda]*.

EXAMINED MATERIAL: **Municipality of Barra Bonita**, Barra Bonita Reservoir, 27-VI-1973, *R. Roque* (SP130785). **Municipality of Barretos**, Barretos, in the city, lake region, with grasses and Cyperaceae, 28-11-1990, *L.H.Z. Branco* (SP255772). **Municipality of Guaratingueta**, Clube dos 500, lake, 01-IV-1966, *C.E.M. Bicudo* (SP96965). **Municipality of Juquiá**, BR-116, km 165, lake, 01-III-1973, *C.EM. Bicudo* & L. *Sormus* (SP113664). **Municipality of Miracatu**, Jaragatiá, Pettena farm, lake, 01-III-1973, *C.E.M. Bicudo, C.R. Leite* & *L. Sormus* (SP113673). **Municipio** *de* **Mirassol**, SP-31, km 410.7, pond with grasses, clay bottom, periphyton, 10-IV-1990, *A.A.J. de Castro* & *C.E.M. Bicudo* (SP255728). **Municipio de Pindamonhangaba**, BR-116, km 270-271, lagoon, 21-V-1966, *C.E.M. Bicudo* (SP96949). **Municipality of Pirassununga**, SP-225, km 23, Cascalho neighborhood, lake, ?-XII-1973, *P.A.C. Senna* (SP123900). **Municipality of Rio Claro**, SP-310, km 156, flooded, 10-V-1973, *C.E.M. Bicudo* & *P.A.C. Senna* (SP104728). **Municipality of Santo Anastácio**, Rodovia Santo Anastácio-Mirante do Paranapanema, Santo Anastácio river, 23- VIII-1973, *D.M. Vital* (SP114511). **Municipality of Santo André**, Estagao Biológica do Alto da Serra de Paranapiacaba, lake near the headquarters, 20-I-1966, *C.E.M. Bicudo* (SP96909). **Municipio de Sao Paulo**, Santo Amaro, lagoon near Av. Washington Luiz n^0 3669, 01- VI-1967, *B. Skvortzov* (SP104098); Parque Estadual das Fontes do Ipiranga, Lago das Gargas, 12-VII-2006, *M. Borduqui,* 23°38'08"S and 23°40'18"S, 46°36'48"W and 46°38'00'W (SP399781). 17-I-2007, M. *Borduqui,* 23°38'08"S and 23°40'18"S, 46°36'48"W and 46°38'00'W (SP399782); Lago das Ninféias, 03-VIII-2007, *T.R. Santos,* 23°38'08"S and 23°40'18"S, 46°36'48"W and 46°38'00"W (SP399783). **Municipality of Tremembé**, SP-442, 13.2 km before Taubaté,

lake on the right, direction Pindamonhangaba-Taubaté, with *Utricularia, Typha* and Cyperaceae, periphyton, 24-IV-1990, *A.A.J. de Castro & C.E.M. Bicudo* (SP255731).
Municipality of Tupa, no precise location, 20-VII-1973, *D.M. Vital* (SP130789).

Comment

Desmodesmus communis (Hegewald) Hegewald is known worldwide as *Scenedesmus quadricauda* (Turpin) Brébisson *sensu* Chodat. Quite variable characteristics in this species are the cell and spine dimensions.

This is one of the most common and widely distributed species in the world under the combination S. *quadricauda* (Turpin) Brébisson *sensu* CHODAT (1913, 1926), which was replaced by HEGEWALD (1977) with *Scenedesmus communis* Hegewald and later by HEGEWALD (2000) with *Desmodesmus communis* (Hegewald) Hegewald.

CARDOSO (1979) discussed the morphological variation he observed in population samples of S. *quadricauda* (Turpin) Brébisson collected in Sao José dos Campos, stating that the metric limits sometimes coincide with and sometimes exceed the limits proposed by UHERKOVICH (1966). However, the size of the spines is not, on its own, a character of sufficient weight to define species in the genus. This characteristic must be associated with others in order to define species, varieties and taxonomic forms of .S' *quadricauda* (Turpin) Brébisson. SMITH (1916) considered S. *quadricauda* (Turpin) Brébisson var. *quadricauda, S. quadrispina* (Chodat) G.M. Smith, S. *longispina* Chodat and S. *nanus* Chodat to be varieties rather than species. Uherkovich (1966) kept the varieties S. *quadricauda* (Turpin) Bréb. var. *quadrispina* (Chodat) G.M. Smith and S. *quadricauda* (Turpin) Brébisson var. *longispina* (Chodat) G.M. Smith, but considered S. *nanus* Chodat to be a distinct species. SANT'ANNA (1984) considered S. *quadricauda* (Turpin) Brébisson to be a species morphologically close to S. *protuberans* Fritsch and S. *opoliensis* Richter, from which it differs in that the cells that make up the cenobium are equally long and do not have prominent cell poles. In particular, we agree with UHERKOVICH (1966) when he mentioned that the variation in shape, size, arrangement of cells in the cenobium, number and pattern and distribution of spines, as well as the poles, are fundamental characteristics in the identification of *Scenedesmus* species.

Desmodesmus communis (Hegewald) Hegewald is morphologically very similar to *D. maximus* (West & West) Hegewald, but the latter has larger cell dimensions.

HEGEWALD (2000) proposed the new combination *Desmodesmus communis* (Hegewald) Hegewald and included most of the taxonomic varieties of S. *quadricauda* (Turpin) Brébisson, which has remained unchanged to this day.

Representatives of *Desmodesmus communis* (Hegewald) Hegewald have been recorded in 14 municipalities in the state of São Paulo. The species is considered to be cosmopolitan worldwide. Despite being easily found, the species is extremely variable in terms of the size and number of spines, cell dimensions and the number of cells in the cenobium, which makes its taxonomic identification somewhat complicated.

Desmodesmus denticulatus* (Lagerheim) An, Friedl & Hegewald var. *denticulatus (Fig. 64)
Algological Studies 96: 9. 2000.
Basionimo: *Scenedesmus denticulatus* Lagerheim, Ofversigt af Kungliga
   Vetenskapsakademiens forhandlingar 39(2): 61, pl. 2, fig. 13-16. 1882.
Synonym: *Scenedesmus arcuatus* Lemmermann f. *spinosus* Hortobágyi &Németh

Cenobium flat, formed by 4 cells; cells arranged alternately, oblong, asymmetrical, poles rounded, outer and inner cells with 1-3 small spines at the cell poles; cell length 10.4-14.7 pm, width 5.5-7.6 pm, polar spines 23 pm long; parietal chloroplastid, 1 pyrenoid.

Geographical distribution in the state of Sao Paulo

IN LITERATURE: **Municipality of Sao Bernardo**, Sao Paulo (SANT'ANNA 1984: 217, fig. 141-142, as *Scenedesmus denticulatus* Lagerheim var. *denticulatus* f. denticulatus*)*. **Municipality of Juquiá** (SANT'*AnnaETAL.* 1988: 91, fig. 41, as *Scenedesmus denticulatus* Lagerheim). **Municipality of Ribeirao Preto** (SILVA 1999: 291, fig. 61, as *Scenedesmus denticulatus* Lagerheim). **Municipality of Sao Paulo** (SANT'ANNA *ETAL.* 1989: 96, fig. 68-69, as *Scenedesmus arcuatus* Lemmermann f. *spinosus* Hortobágyi & Németh; SANT'ANNA *ET AL.* 1989: 96, fig. 84-85, as *Scenedesmus denticulatus* Lagerheim). **Municipio de Sumaré** (SANT'ANNA 1984: 203, fig. 129, as *Scenedesmus arcuatus* Lemmermann f. *spinosus* Hortobágyi & Németh).

EXAMINED MATERIAL: **Municipality of Pilar do Sul**, SP-250, km 127, to the right, direction Sao Paulo-Pilar do Sul, Turvinho neighborhood, Turvinho river, periphyton among grasses, 17- IV-1989, *A.A.J. de Castro, C.E.M. Bicudo & D.C. Bicudo* (SP188431). **Municipio de Sao Paulo**, Parque Estadual das Fontes do Ipiranga, Lago das Garbas, col. *M. Borduqui,* 17-I- 2007. GPS 23°38'08"S and 23°40'18"S

and 46°36'48"W and 46°38'00"W (SP399782).
Comment

Desmodesmus denticulatus (Lagerheim) An *et al.* is a well-defined species that is easy to identify taxonomically. According to COMAS (1996), there are characteristics of this species that are very striking, namely: the four alternating cells are all the same, with one to four denticles oriented parallel to each other or slightly inclined in relation to the cenobium. *Scenedesmus denticulatus* (Lagerheim) An *et al.* is common and well represented in samples from the state of Sao Paulo.

SANT'ANNA (1984) mentioned having observed individuals of the *denticulatus* type alongside others of the *smithii type;* however, the only difference between these two types is the presence or absence of an angularity in the region of contact between the cells. The author also mentioned that these populations are probably representatives of the same species and that the difference may be due to environmental factors in each location. We agree with SANT'ANNA (1984) when he stated that S. *smithii sensu* PHILIPOSE (1967) is a heterotypic synonym of S. *denticulatus* G.M. Smith. KOMÁREK (1983) and KOMÁREK & FOTT (1983) differentiated S. *smithii* Teiling from S. *denticulatus* G.M. Smith by the elongated oval cells and the formation of cenobia in two rows, which are characteristics of S. *denticulatus* G.M. Smith. HINDÁK (1990) did not accept the separation of these two species, considering S. *smithii* Teiling to be synonymous with S. *denticulatus* G.M. Smith. According to COMAS (1996), for both morphological types to be considered representatives of independent species, it is necessary to assign another name to the expression *smithii,* since it is a later homonym of S. *smithii* Chodat. HEGEWALD (2000) considered S. *Smith* Teiling to be a synonym of *D. denticulatus* G.M. Smith.

Representatives of D. *denticulatus* (Lagerheim) An *et al.* var. *denticulatus* have been recorded in only two localities in the state of São Paulo, but the species is considered to be easy to identify taxonomically. We consider it important to remember that D. *denticulatus* (Lagerheim) An *et al.* var. *denticulatus* is morphologically very similar to *Desmodesmus spinulatus* (Biswas) Hegewald, but they differ in the shape of the cell, which in the former is oblong and asymmetrical and in the latter is elliptical-fusiform, with truncated poles. Small populations of D. *denticulatus* (Lagerheim) An *et al.* var. *denticulatus* were found, however, their constituent specimens showed all the diagnostic characteristics of the species.

***Desmodesmus denticulatus* (Lagerheim) An, Friedl & Hegewald var. *linearis* (Hansgirg) Hegewald**
(Fig. 65-68)
Algological Studies 96: 10. 2000.
Basionym: *Scenedesmus denticulatus* Lagerheim var. *linearis* Hansgirg, Prodromus der
           Algenflora von Böhmen 1: 268. 1888.
Synonyms: *Scenedesmus brevispina* (G.M. Smith) Chodat
            Scenedesmus denticulatus Lagerheim var. *australis* Playfair

Flat cenobium formed by 4-8 cells; cells arranged linearly, oblong cells, rounded poles, outer cells well convex in the median part, outer and inner cells with small spines; cell length 5.3-20.0 pm, width 2.28.0 pm, polar spines 1-3 pm long; parietal chloroplastid, 1 pyrenoid.
Geographical distribution in the State of Sao Paulo

IN LITERATURE: **Municipality of Bofete** and **Jaú** (SANT'ANNA 1984: 219, fig. 145, as *Scenedesmus denticulatus* Lagerheim var. *australis* Playfair). **Municipality of Sao José dos Campos** (CARDOSO 1979: 84, fig. 104-105, as *Scenedesmus denticulatus* Lagerheim var. *linearis* Playfair). **Municipality of Sao Paulo** (SANT'ANNA *etAL.* 1989: 96, fig. 79, as *Scenedesmus brevispina* (G.M. Smith) Chodat; SANT'ANNA *ET AL.* 1989: 96, fig. 86, as *Scenedesmus denticulatus* Lagerheim var. *australis* Playfair; FERRAGUT *ETAL.* 2005: 153, fig. 77, such as *Desmodesmus denticulatus* (Lagerheim) An *et al.* var. *linearis* (Hansgirg) Hegewald).

EXAMINED MATERIAL: **Municipality of Casa Branca**, SP-340, km 228.5, swamp effluent stream, right side, 17-X-1989, *A.A.J de Castro & C.E.M. Bicudo* (SP188321). **Municipality of Miracatu**, Jaracatiá, Pettena farm, lake, 01-III-1973, *C.E.M. Bicudo, C.R. Leite & L. Sormus* (SP113673). **Municipality of Pindamonhangaba**, Sao loao farm, Sao loao lake, 21-V-1966, *C.E.M. Bicudo* (SP96950). **Municipality of Pirassununga**, SP-225, km 23, Cascalho neighborhood, lake, ?-XII-1973, *P.A.C. Senna* (SP123900). **Municipality of Rio Claro**, SP-310, km 156, flooded, 10-V-1973, *C.E.M. Bicudo & P.A.C. Senna* (SP104728). **Municipio de Sao Carlos**, SP-310, km 222, aldeia Conde do Pinhal, stream, 10-V-1973, *C.E.M. Bicudo & L. Sormus* (SP104723). **Municipality of Sao Paulo**, Santo Amaro, pond near Av. Washington Luiz n^0 3669, 01-VI-1967, *B. Skvortzov* (SP104098). **Municipality of Sao Paulo**, Parque Estadual das Fontes do Ipiranga, Lago das Ninféias, 03-VIII-2007, *T.R. Santos,* 23°38'08"S and 23°40'18"S, 46°36'48'W and 46°38'00'W (SP399783).

Comment

Desmodesmus denticulatus (Lagerheim) An *et al.* var. *linearis* (Hansgirg) Hegewald originally has, according to NOGUEIRA (1991), a linear arrangement of the cells in the cenobium, which differs from representatives of the typical variety of the species, which have an alternating arrangement. The material examined here has two small spines on the poles, which is also a difference from the typical variety of the species, which has one to four small spines on each cell pole (SANT'ANNA 1984).

KOMÁREK & FOTT (1983) questioned the real existence of this variety, as HANSGIRG (1888) did not provide an illustration and the type material is unknown. This name has, however, been mentioned frequently. However, it is usually *Scenedesmus aculeolatus* Reinsch, a species that has not been transferred to *Desmodesmus*. HEGEWALD (1985) selected a culture from Peru (HEGEWALD 1977) as the neotype of S. *aculeolatus* Reinsch. *Scenedesmus aculeolatus* Reinsch has also been found in India and Cuba. The latter species is typical for its oval-cylindrical cells joined along almost their entire length to form linear, rarely slightly alternating cenobia surrounded by mucilage. There is usually a denticle at each cell pole and, optionally, other denticles on the cell surface.

SANT'ANNA *ET AL.* (1989) and FERRAGUT *ET AL.* (2005) documented the presence of representatives of D. *denticulatus* (Lagerheim) An *et al.* var. *linearis* (Hansgirg) Hegewald with two denticles in each of the two cells of the cenobium. No specimens of this type were found in the populations studied here. We considered the specimens in SANT'ANNA *ET AL.* (1989) and FERRAGUT *etAL.* (2005) as an expression of the spectrum of morphological variation of the current var. *linearis* (Hansgirg) Hegewald. PHILIPOSE (1967) considered it normal for *Scenedesmus denticulatus* Lagerheim var. *linearis* Hansgirg (today *Desmodesmus denticulatus* (Lagerheim) An *et al.* var. *linearis* (Hansgirg) Hegewald) to have a single denticle per cell, but two spines can occur (rare), as has happened in populations in the state of São Paulo (SANT'ANNA ET *al.* 1989, FERRAGUT ET *al.* 2005).

Representatives of D. *denticulatus* (Lagerheim) An *et al.* var. *linearis* (Hansgirg) Hegewald were found in seven localities in the state of Sao Paulo and are easy to identify taxonomically.

Desmodesmus dispar (Brébisson) Hegewald (Fig. 69-70)

Algological Studies 96: 10. 2000.

Basiönimo: *Scenedesmus dispar* **Brébisson**, Mémoires de la Société Imperiale des Sciences Naturelles de Cherbourg 4: 159, pl. I, fig. 32. 1856.

Synonym: *Scenedesmus longus* Meyen var. *dispar* (Brébisson.) G.M. Smith

Flat cenobium formed by 2-4 cells; cells arranged linearly, elliptical, poles rounded; outer cells with 1 medium spine arranged diagonally opposite each other at one pole, 1 small straight spine also arranged diagonally opposite each other at the other pole, inner cells with small curved spines, facing downwards or straight; cell length 8.4-13.9 pm, width 2.6-5.9 pm, polar spines 2-10 pm long; parietal chloroplastid, 1 pyrenoid.

Geographical distribution in the State of Sao Paulo

IN LITERATURE: **Municipio de Sao Paulo** (*FerragutetAL.* 2005: 155, fig. 91, as *Scenedesmus danubialis* Hortobágyi).

EXAMINED MATERIAL: **Municipality of Barretos**, Barretos, in the city, lake region, with grasses and Cyperaceae, 28-II-1990, *L.H.Z. Branco* (SP255772). **Municipality of Capao Bonito**, SP-127, km 199.9, river, 18-VII-2000, *C.E.M. Bicudo, F.C. Pereira & L.L. Morandi* (SP365693). **Municipio de Inúbia Paulista**, SP-294, km 578, stream with vegetation almost completely covering it, periphyton and phytoplankton, 20-VII-1991, *M.C. Bittencourt-Oliveira* (SP239091). **Municipio de Sao Paulo**, Parque Estadual das Fontes do Ipiranga, Lago do IAG, col. *I.S. Vercellino,* 03-VIII-1998, 23° 38'08"S e 23° 40'18"S, 46°36'48"W e 46°38'00"W (SP399779); 06-III-1999, 23°38'08"S e 23°40'18"S, 46°36'48"W e 46°38'00"W (SP399780); Lago das Ninféias, col. *TR. Santos,* 03-VIII-2007, 23°38'08"S and 23°40'18"S, 46°36'48"W and 46°38'00"W (SP399783).

Comment

Desmodesmus dispar (Brébisson) Hegewald has a very eloquent diagnostic feature, which is THE POSITIONING OF THE POLAR SPINES IN THE OUTSIDE CELLS with a medium-sized thorn located diagonally opposite each other at one of the poles and a small thorn also located diagonally opposite each other at the other pole. In addition to this characteristic, there are also spines on the inner cells which can also be extremely variable, ranging from straight and parallel to the longitunal axis to convex or concave, as well as rosettes which can occur on the outer cells of the cenobium.

FERRAGUT *ET AL.* (2005) identified material collected from the IAG Lake as *Scenedesmus danubialis* Hortobágyi. The re-examination of this material, however, led to the conclusion that it was D. *dispar* (Brébisson) Hegewald.

Representatives of D. *dispar* (Brébisson) Hegewald have been recorded at four sites in the state of São Paulo, always forming small populations of just a few individuals. However, the species is easily identified, except for the bicaudate forms which can be confused with representatives of *Desmodesmus armatus* (Chodat) Hegewald var. *bicaudatus* (Guglielmetti) Hegewald. The two forms do not differ greatly, except for the small spines on the inner cells that may or may not occur. The only way to distinguish the morphological expressions of D. *armatus* (Chodat) Hegewald var. *bicaudatus* (Guglielmetti) Hegewald from those of D. *dispar* (Brébisson) Hegewald is by analyzing populations.

Desmodesmus flavescens (R. Chodat) Hegewald var. *breviaculeatus* (Bourrelly) Hegewald (Fig. 71-72)

Algological Studies 96: 10. 2000.

Basionym: *Scenedesmus tenuispina* Chodat var. *breviaculeatus* Bourrelly, Bibliotheca Phycologica 76: 57, pl. 23, fig. 7-9. 1987.

Synonym: *Scenedesmus courbetensis* Therézien & Coute

Flat cenobium formed by 2 cells; cells arranged linearly, elliptical, poles rounded, 1-3 small spines at the poles, 1 pair of equatorial spines in the median part of the cell; cell length 11.2-13.0 pm, width 3.2-7.0 pm, polar spines 2-4 pm long; parietal chloroplastidium, 1 pyrenoid.

Geographical distribution in the State of Sao Paulo

IN LITERATURE: **Municipality of Juquiá** (SANT'ANNA *ET AL.* 1988: 91, fig. 42, as *Scenedesmus gutwinskii* Chodat). **Municipality of Sao Paulo** (FERRAGUT *ET AL.* 2005: 155, fig. 94, as *Scenedesmus gutwinskii* Chodat).

EXAMINED MATERIAL: **Municipality of Sao Luis do Paraitinga**, SP-125, km 34.7, school on the right, direction Taubaté-Sao Luís do Paraitinga, drain behind school, in the middle of Cyperaceeae, *Typha* and Liliaceae, 27-XI-1989, *A.A.J. de Castro & C.E.M. Bicudo & E.M. De-Lamonica-Freire* (SP188323). **Municipio de Sao Paulo**, Parque Estadual das Fontes do Ipiranga, Lago do IAG, 06-III-1999, *I.S. Vercellino,* 23° 38'08"S and 23° 40'18"S, 46°36'48"W and 46°38'00"W (SP399780); Lago das Garbas, 17-I-2007, M. *Borduqui,* 23°38'08"S and 23°40'18"S, 46°36'48"W and 46°38'00"W (SP399782).

Comment

KOMÁREK & FOTT (1983) mentioned only four-celled cenobia, however, in the samples examined here, only individuals with two cells were seen. However, we agree with the above authors when they state that the existence of two equatorial lateral spines on the outer cell is the diagnostic characteristic of the species.

The species is characterized by the presence of two equatorial lateral spines on the marginal cells, not arranged longitudinally and each inserted laterally in the median region of the marginal cell. The material currently studied corresponds to var. *breviaculeatus* (Bourrelly) Hegewald, especially with the iconotype of *Scenedesmus courbetensis* Therézien & Couté *nom. ileg.* (HEGEWALD & SILVA 1988: 178-179, fig. 282). In addition, the species has one or two small polar spines on the cells of the cenobium.

It should be mentioned that D. *flavescens* (R. Chodat) Hegewald var. *breviaculeatus* (Bourrelly) Hegewald. is very similar to S. *gutwinskii* Chodat and that the two species can easily be confused; however, the latter species does not have equatorial spines but lateral ones.

HEGEWALD (2000) considered *Scenedesmus gutwinskii* Chodat var. *heterospina* Bodrogkozy and *Scenedesmus gutwinskii* Chodat f. *natrophilus* Kiss to be synonyms of D. *subspicatus* (R. Chodat) Hegewald & A. Schmidt, but not the typical variety of S. *gutwinskii* Chodat. *Scenedesmus gutwinskii* Chodat var. *heterospina* Bodrogkozy is given a new name and species category by CHODAT (1916) based on *Scenedesmus quadricauda* (Turpin) Brébisson f. *hyperabundans* Gutwinski, which is described again and has its first illustration published (GUTWINSKI 1891). According to this figure (see HEGEWALD & SILVA 1988: 447-448, fig. 717), the specimens of this variety have three equatorial lateral spines on the marginal cells, a fact that closely relates them to D. *flavescens* (R. Chodat) Hegewald.

Desmodesmus flavescens (R. Chodat) Hegewald var. *breviaculeatus* (Bourrelly) Hegewald differs from *D. abundans* (Kirchner) Chodat by having equatorial lateral spines. *Desmodesmus flavescens* (R. Chodat) Hegewald var. *breviaculeatus* (Bourrelly) Hegewald differs from the typical variety of the species due to the size of the spines, which in the latter are much larger than in var. *breviaculeatus* (Bourrelly) Hegewald.

FERRAGUT *ET AL.* (2005) identified *Scenedesmus gutwinskii* Chodat from material collected in Lago do lAG, in the Municipality of São Paulo, however, when re-identifying this material we considered it to be identical to D. *flavescens* (R. Chodat) Hegewald var. *breviaculeatus* (Bourrelly) Hegewald. The same happened with the material in SANT'ANNA *ET AL.* (1988) identified as *Scenedesmus gutwinskii* Chodat, however, it is also D. *flavescens* (R. Chodat) Hegewald var. *breviaculeatus* (Bourrelly) Hegewald.

HEGEWALD (1993) proposed the new combination *Desmodesmus flavescens* (R. Chodat) Hegewald var. *breviaculeatus* (Bourrelly) Hegewald and mentioned all the morphological variability of this var. *breviaculeatus* (Bourrely) Hegewald in culture, mentioning the rare occurrence of cenobia with only one equatorial thorn.

Representatives of D. *flavescens* (R. Chodat) Hegewald var. *breviaculeatus* (Bourrelly) Hegewald were detected in three localities in the state of São Paulo. However, the material is easy to identify taxonomically as it has equatorial spines in the median part of the cell. The species showed no significant morphological variation in the few individuals that made up each population examined.

***Scenedesmus gutwinskii* Chodat var. *bekesensis* Uherkovich,** name not validly published (Fig. 73-74)
Die *Scenedesmus-Arten* Ungarns. 111, fig. 786-787. 1966.

Flat cenobium formed by 4 cells; cells arranged linearly, elliptical, poles rounded; outer cells with medium spines at the poles, small spines along the entire length of the margin, inner cells with or without 1 small spine; cell length 6.2-7.5 pm, width 1.2-2.0 pm, polar spines 1-6 pm long; parietal chloroplastidium, 1 pyrenoid.

Geographical distribution in the State of Sao Paulo

IN LITERATURE: nothing. First mention of the occurrence of this variety in the state of Sao Paulo.

EXAMINED MATERIAL: **Municipio de Sao Paulo**, Horto Florestal, lake, 16-VII- 1962, *C.E.M. Bicudo &R.M.T. Bicudo* (SP96850).

Comment

Scenedesmus gutwinskii Chodat var. *bekesensis* Uherkovich is a name that has not been validly published because it lacks a diagnosis and/or description in Latin (art. 36 of the International Code of Botanical Nomenclature) and a nomenclatural type designation (art. 37 of the International Code of Botanical Nomenclature). In our opinion, it should be included in the synonymy of *D. subspicatus* (R. Chodat) Hegewald & Schmidt, but as an independent variety. The proposal to be considered requests, preferably, the assignment of a new name.

The species has only been found in one municipality in the state of Sao Paulo, namely in the city of Sao Paulo, in a lake located in the Horto Florestal. The sample examined consisted of only four individuals, which is why we consider the species to be a rare occurrence in the state. With regard to morphological variability, the only character that varied was the presence of small spines at the poles of the inner cells; all the other characters remained stable within the small population examined.

***Desmodesmus intermedias* (R. Chodat) Hegewald var. *intermedias* (Fig. 75)**
Algological Studies 96: 11. 2000.

Basionimo: *Scenedesmus intermedias* R. Chodat, Zeitschrift fur Hydrologie 3: 231, fig. 135. 1926.

Cenobium planes formed by 4 cells; cells arranged alternately, elliptic- oblong, poles rounded, outer cells with long spines at the poles, inner cells with small spines; cell length 4.5-12.8 pm, width 3.0-5.4 pm, polar spines 3.5-13.0 pm long; parietal chloroplastidium, 1 pyrenoid.

Geographical distribution in the State of Sao Paulo

IN LITERATURE: **Municipality of Sao Paulo** (SANT'ANNA *ET AL.* 1989: 97, fig. 87, as *Scenedesmus intermedius* Chodat; FERRAGUT *ETAL.* 2005: 153, fig. 79).

EXAMINED MATERIAL: **Municipality of Barretos**, Barretos, in the city, lake region, with grasses and Cyperaceae, 28-II-1990, LH.Z. *Branco,* (SP255772).

Comment

Desmodesmas intermedias (R. Chodat) Hegewald var. *intermedias* was previously identified as *Scenedesmas sooi* Hortobágyi according to UHERKOVICH (1966), however, according to HEGEWALD (2000), the latter species is synonymous with *Desmodesmus intermedias* (R. Chodat) Hegewald var. *intermedias*.

HEGEWALD *ETAL.* (1998) stated that .S' *intermedias* is a very variable species, and can have oval or elongated cells, arranged linearly or alternately in the cenobium, with or without ribs, with or without spines, bicaudate or quadricaudate and with or without spines in the inner cells. What characterizes this species is the ultrastructure of the wall.

ÑIn the samples currently analyzed, D. *intermedias* (R. Chodat) Hegewald var. *intermedias* showed cenobia formed by alternately distributed ovoid cells. Specimens of this variety have only been found in one locality in the state of Sao Paulo, and mainly due to the fact that few individuals of this species have been found. However, its identification was not considered a problem due to the shape of the cells and their alternating arrangement in the cenobium. The specimens analyzed here did not show polar spines on the inner cells, although HUSZAR (1985) found that specimens from the state of Rio de Janeiro showed spines on the cell poles of the inner cells. However, SAÑT'AÑÑA *etAL.* (1989) and ÑOGUEIRA (1991) did not see spines on the inner cells, which shows that the presence or absence of spines on the inner cells is a considerable morphological variation within the species. HUSZAR (1985) and FERRAGUT *ET AL.* (2005) did, however, study individuals with spines on the inner cells.

Desmodesmus intermedias (R. Chodat) Hegewald var. *acutispinus* **(Roll) Hegewald** (Fig. 76-81) Algological Studies 96: 12. 2000.

Basionimo: *Scenedesmus quadricauda* **var.** *acutispinus* **Roll**, Russkii arkhiv protistologii 4: 144, 149.1925.

Synonym: *Scenedesmas bicaadatas* Dedusenko

Flat cenobium formed by 2-8 cells; cells arranged linearly, elliptical to oblong, poles rounded, outer cells with 1 spine on only one of the poles, arranged in opposite directions (diagonally), inner cells with or without small spines, median longitudinal ribs sometimes present; cell length 6.2-12.8 pin, width 1.9-5.3 pm, polar spines 3-14 pm long; parietal chloroplastidium, 1 pyrenoid.

Geographical distribution in the State of Sao Paulo

IN LITERATURE: **Municipalities of Americana, Atibaia, Barra Bonita, Guaratinguetá, Moji das Cruzes, Pindamonhangaba, Registro, Santo André, Sao Bernardo do Campo** and **Tupa** [SANT'ANNA 1984: 204, fig. 130, as *Scenedesmus bicaudatus* (Hansgirg) Chodat]. **Municipality of Sao Carlos** [SCHWARZBOLD 1992: 112, fig. 4, as *Scenedesmus bicaudatus* (Hansgirg) Chodat]. **Municipality of Sao José dos Campos** (CARDOSO 1979: 81, fig. 106-107, as *Scenedesmus bicaudatus* (Hansgirg) Chodat var. *bicaudatus).* **Municipio de Sao Paulo** [SANT'ANNA *ET AL.* 1989: 96, fig. 76, as *Scenedesmus bicaudatus* Dedusenko; GENTIL 2000: 55, fig. 14, as *Scenedesmus bicaudatus* Dedusenko; *FERRAGUTETAL.* 2005: 153, fig. 78, as *Desmodesmus intermedias* (R. Chodat) Hegewald var. *acutispinus* (Roll) Hegewald].

EXAMINED MATERIAL: **Municipality of Angatuba**, SP-270, km 203.7, on the right towards Angatuba-Itapetininga, Casa da Pedra farm, pond with Cyperaceae, 17-IV-1989, *A.A.J. de Castro & C.E.M. Bicudo* (SP188215). **Ara^atuba municipality,** Marechal Rondon highway, unspecified location, with Cyperaceae, grasses and *Myriophyllum,* 15-I-1992, *L.H.Z. Branco* (SP239239). **Municipality of Arujá**, Clube Fiscal do Brasil, lake, 01-V-1975, *L. Sormus* (SP130815). **Municipality of Itapura**, Tiete river, SP-595, km 21.5, 16-V-2001, *C.EM. Bicudo* & D.C. *Bicudo,* 22°16'41.0" S, 51°48'16.5" W, conductivity 40 pS.cm^{-1} , pH 6.0 (SP355388). **Municipio de Juquiá**, BR-116, km 165, lake, 01-III-1973, *C.EM. Bicudo* &L. *Sormus* (SP113664). **Municipality of Pindamonhangaba,** BR-116, km 270-271, lake, 21-V-1966, *C.E.M. Bicudo* (SP96949). **Municipality of Pirassununga**, SP-225, km 23, Cascalho neighborhood, lake, ?-XII-1973, *P.A.C. Senna* (SP123900). **Municipality of Tupa**, no precise location, 20-VII-1973, *D.M. Vital* (SP130789). **Municipality of Sao Miguel Arcanjo**, SP-250, left side, 200 m, direction Sao Miguel-Itapetininga, dam formed by the acude stream, with water hyacinth, 17-IV-1989, *A.A.J. de Castro & C.E.M. Bicudo & D.C. Bicudo* (SP188214). **Municipio de Sao Paulo**, Parque Estadual das Fontes do Ipiranga, Lago das Gargas, 12-VII-2006, col. M. *Borduqui,* 23°38'08"S and 23°40'18"S, 46°36'48"W and 46°38'00"W (SP399781); 17-I-2007, col. *M. Borduqui,* 23°38'08"S and 23°40'18"S, 46°36'48"W and 46°38'00"W (SP399782); Lago das Ninféias, 03-VIII-2007, *T.R. Santos,* 23°38'08"S and 23°40'18"S, 46°36'48"W and 46°38'00"W (SP399783).

Comment

Desmodesmus intermedias (R. Chodat) Hegewald var. *acutispinus* (Roll) Hegewald is well represented in the State of São Paulo, having been collected in 10 localities and easily found in the samples analyzed. It is also a species that is easy to identify taxonomically because of the two inclined, opposite spines, arranged diagonally in the outer cells of the cenobium. These spines can vary in size from one individual to another. When the spines are very small, specimens with them can be mistaken for *Scenedesmus bicaudatus* (Hansgirg) Chodat var. *brevicaudatus* Hortobágyi. It is possible that such specimens representing this variety are nothing more than a morphological variation of the species and not a taxonomic variety as such.

Desmodesmus intermedius (R. Chodat) Hegewald var. *acutispinus* (Roll) Hegewald is very similar to *Desmodesmus armatus* (R. Chodat) Hegewald var. *bicaudatus (*Guglielmetti) Hegewald, but they differ in that *Desmodesmus armatus* (R. Chodat) Hegewald var. *bicaudatus* (Guglielmetti) Hegewald has tiny spines along the margin of the outer cells, resembling a comb, a structure that does not occur in *Desmodesmus intermedius* (R. Chodat) Hegewald var. *acutispinus* (Roll) Hegewald. In addition, the cenobium in the latter species is made up of alternating cells, while in the former it is linear.

HEGEWALD (2000) included, without any additional information, var. *acutispinus* Roll in *D. intermedius (*R. Chodat) Hegewald, despite the fact that the original figure (see HEGEWALD & SILVA 1988: 430, fig. 688) shows a cenobium with aligned cells and not alternating ones, as is characteristic of *D.*

intermedius (R. Chodat) Hegewald.

SANT'ANNA (1984), SCHWARZBOLD (1992), GENTIL (2000) and FERRAGUT *ETAL.* (2005) observed and illustrated specimens with linear cenobiums, which shows that linear cenobiums and alternating cenobiums are variable within the species, so this characteristic should not be considered a good criterion for defining this species. In addition to the morphological variation of the cenobium, D. *intermedius* (R. Chodat) Hegewald var. *acutispinus* (Roll) Hegewald also shows great variation with regard to the presence or absence of ribs, the presence or absence of small polar spines in the inner cells and with regard to the size of the spines.

According to HEGEWALD *ETAL.* (1998) it is possible to recognize two varieties of this species - var. *acutispinus* (Roll) Hegewald & An and var. *inflatus* (Svirenko) Hegewald & An - in addition to the typical one. These varieties are defined by having only two diagonally arranged spines. Both were transferred to *Desmodesmus: D. intermedius* (R. Chodat) Hegewald var. *acutispinus* (Roll) Hegewald and *D. intermedius* (R. Chodat) Hegewald var. *inflatus* (Svirenko) Hegewald (HEGEWALD 2000). Var. *acutispinus* (Roll) Hegewald is characterized by its oval-long bicaudate cells arranged in a single row.

Desmodesmus lunatus (West & West) Hegewald (Fig. 82-84)
Algological Studies 96: 13. 2000.
Basionym: *Scenedesmus denticulatus* Lagerheim var. *lunatus* West & West, Transactions of the
   Linnean Society of London, Botánica series 5: 83, pl. 5, fig. 11-12. 1895.
Synonyms: *Scenedesmus lunatus* (West & West) R. Chodat
   Scenedesmus polyglobulus Hortobágyi

Flat cenobia formed by 2-4 cells; cells arranged linearly, cylindrical-fusiform, poles truncated to rounded, outer cells slightly arched, outer and inner cells with 1-3 small teeth at the poles; cell length 5.9-10.5 pm, width 1.5-3.45 pm long; polar spines 1-3 pm long; chloroplastidia parietal, 1-pyrenoid.

Geographical distribution in the State of Sao Paulo

IN THE LITERATURE: nothing. First mention of the occurrence of the species in the state of Sao Paulo.

EXAMINED MATERIAL: **Municipality of Campos de Jordao**, Alagoinha neighborhood, lake, 11-III-1973, *M.M. Sakane* (SP130445). **Municipality of Cananéia**, Ilha Comprida, 120 m from the sea, lagoon, 07-III-1975, *M. Vital* (SP130813). **Municipality of Itaporanga**, river, SP-255, km 358, 26-VII-2000, *S.M.M. Faustino, & S.P. Schetty,* 23°42'24.3" S, 49°28'15.6" W, conductivity 20 pS cm^{-1} , pH 7.1 (SP355358). **Municipality of Juquiá**, BR-116, km 160, flooded, 01-III-1973, *C.E.M. Bicudo & L. Sormus* (SP113672). **Municipality of Pirassununga**, SP-225, km 23, Cascalho neighborhood, lake, ?-XII-1973, *P.A.C. Senna* (SP123900). **Municipality of Pitangueiras**, acude, SP-322, km 368, 16-VII-2000, *C.EM. Bicudo, SMM. Faustino, &L.L. Morandi,* 20°59'30.5"S, 48°14'01.1"W, conductivity 40 pS cm^{-1} , pH 6.5 (SP355382). **Municipality of Rancharía**, Vila Agisse, Ribeirao Capivari, 25-VIII-1973, *D.M. Vital* (SP114513). **Municipality of Rio Claro**, Horto Florestal, lake, 31-I-1975, *O.A. da Silva* (SP123867). **Municipality of Sao José do Rio Pardo**, SP-350, km 265, acude, 08-VIII-2000, *C.E.M. Bicudo, L.A. Carneiro & S.M.M. Faustino* (SP365698).

Comment

According to COMAS (1996), *Desmodesmus lunatus* (West & West) Hegewald can be identified by the lunate shape of its cells and the small warty structures, which can also occur in the form of denticles. *Desmodesmus lunatus* (West & West) Hegewald has inner cells with denticles, which are very characteristic of the species. Individuals without denticles can also occur, but the lunate shape of the cell confirms the species' identity. Hence the importance of studying popular species because sometimes there are individuals which, due to intra-population variation, on the one hand are not so characteristic but, on the other hand, still represent the species.

This species was recorded for the first time in the Americas by COMAS (1980). We agree with this author that the material from Cuba resembles that of HORTOBÁGYI (1967), however, the same author in 1984 (COMAS 1984) mentioned that the typical granulations are absent and it is not possible to state whether what was observed were actually granulations or cell wall pores.

HEGEWALD *ET AL.* (1998) studied the ultrastructure of the cell wall of D. *lunatus* (West & West) Hegewald and stated that, regardless of whether it was a natural or cultured sample, the lunate outer cells are the only stable characteristic of this species. The inner cells showed a slight curvature reminiscent of the lunar shape and, with regard to the cenobia, there were sporadic cenobia with alternating cells. HEGEWALD *ET AL.* (1998) also noted the presence of globular warts and other smaller warts over the entire surface of the cell wall, which is typical of the species.

Specimens of this species were collected from nine localities in the state of Sao Paulo. *Desmodesmus lunatus* (West & West) Hegewald is easy to identify because it has lunate cells with denticles or warts at the poles.

Desmodesmus maximus (West & West) Hegewald (Fig. 85-87)

Algological Studies 96: 13. 2000.

Basionimo: *Scenedesmus quadricauda var. maximus* West & West, Transactions of the
Linnean Society of London: Botánica series, 5: 83. 1895.

Synonym: *Scenedesmus quadricauda* (Turpin) Brébisson var. *westii* G.M. Smith

Flat cenobium formed by 2-16 cells; cells arranged linearly, elliptical to oblong, poles rounded, outer cells with 1 spine at each pole, inner cells without spines, outer cells may have sharply curved spines; cell length 13.6-27.8 pm, width 4.6-8.9 pm, polar spines 5-25 pm long; parietal chloroplastidium, 1 pyrenoid.

Geographical distribution in the state of Sao Paulo

IN LITERATURE: **Municipios de Campos do Jordao, Guaratinguetá, Juquiá, Sao Carlos, Sao Paulo** [SANT'ANNA 1984: 242, fig. 161, as *Scenedesmus quadricauda* (Turpin) Brébisson var. *maximus* West & West]. **Municipios de Barra Bonita, Juquiá e Sao Paulo** [SANT'ANNA 1984: 248, fig. 162-164, as *Scenedesmus quadricauda* (Turpin) Brébisson var. *westii* G.M. Smith]. **Municipio de Sao Paulo** [FERRAGUT *etAL.* 2005: 153, fig. 80, as *Desmodesmus maximus* (West & West) Hegewald].

EXAMINED MATERIAL: **Municipality of Campos do Jordao**, Alagoinha neighborhood, lake, 11-III-1973, *M.M. Sakane* (SP130445). **Municipality of Ibirá**, Termas de Ibirá, lake, 25- V-1973, *D.M. Vital* (SP113429). **Municipality of Juquiá**, BR-116, km 165, lake, 01-III-1973, *C.EM. Bicudo &L. Sormus* (SP113664). **Municipality of Tambaú**, Clube de Tambaú, dam, 23-VI-1973, *D.M. Vital* (SP113574).

Comment

Cenobia with widely varying cell sizes and spines. The cenobia are linear and made up of oblong cells with well-rounded poles and a large, long spine on the outer cells. The outer cells may have sharply curved spines, which are used as the diagnostic feature of *Scenedesmus quadricauda* (Turpin) Brébisson var. *westii* G.M. Smith. This variety was considered by HEGEWALD (2000) to be a morphological variation of *Desmodesmus maximus* (West & West) Hegewald. There are no denticles or warts on the outer cells and no spines on the inner cells of the cenobium. HEGEWALD (2000) considered 32 synonyms of *D. maximus* (West & West) Hegewald, of which the following were recorded for the state of Sao Paulo: *Scenedesmus quadricauda* (Turpin) Brébisson var. *maximus* West & West and S. *quadricauda* (Turpin) Brébisson var. *westii* G.M. Smith.

Desmodesmus maximus (West & West) Hegewald can be confused with *Desmodesmus communis* (Hegewald) Hegewald, but differs in the size of the cells, which are much larger in the first species.

Specimens of D. *maximus* (West & West) Hegewald were collected in four localities in the state of Sao Paulo and the species can be considered a common occurrence in the state. However, care should be taken when identifying it so as not to confuse it with D. *communis* (Hegewald) Hegewald.

Desmodesmus opoliensis (P. Richter) Hegewald var. *opoliensis* (Fig. 88)

Algological Studies 96: 14. 2000.

Basionimo: *Scenedesmus opoliensis* P. Richter, Zeitschrift angewandte Mikroskopie und Klinische Chemie 1: 3, 7, fig. a-e. 1895.

Cenobium formed by 2-4 cells; cells arranged linearly or alternately, elliptical to fusiform, poles acute, truncated, outer cells with 1 protuberance from which 1 wide, eccentric spine emerges; cell wall smooth, inner cells in contact only by the subapical part, with or without small spines at the poles; cell length 8.0-28.0 qm, width 2.5-9.0 qm, spine 10-28 qm long.

Geographical distribution in the State of Sao Paulo

IN LITERATURE: **Municipalities of Jaú, Rio Claro, Santo André** and **Sao Bernardo do Campo** (SANT'ANNA 1984: 228, fig. 152, as *Scenedesmus opoliensis* P. Richter). **Municipality of Sao Paulo** (SANT'ANNA *ETAL.* 1989: 97, fig. 90, as *Scenedesmus opoliensis* P. Richter).

MATERIAL EXAMINED: only material from the literature.

Comment

Desmodesmus opoliensis (P. Richter) Hegewald has cells that range from ellipsoid to fusiform. According to COMAS (1996), the ends are attenuated and the poles are more or less truncated in the outer cells; the outer cells are arched and the inner cells are straight. The spines are inserted at the angles of the poles of the outer cells of the cenobium. The diagnostic feature of D. *opoliensis* (P. Richter) Hegewald is these spines located eccentrically on the outer cells of the cenobium. This feature differentiates this species from others, such as *Desmodesmus protuberans* (Fritsch & Rich) Hegewald. Another important characteristic of the typical variety of the species is the formation of a type of cenobium in which the cells can be arranged linearly or alternately, but without contacting each other completely, but only in the region close to the poles (subapically), which differentiates the typical variety from var. *mononensis* (R. Chodat) Hegewald, which has the cenobium cells entirely in contact.

According to the most recent criteria, this is a species with a wide circumscription (HINDÁK 1990, HEGEWALD 2000), one of whose most important diagnostic features is the truncated poles of the marginal cells, where the spines, which are not always very developed at the pole angles, are inserted. The typical variety of the species is characterized by its markedly alternate cells. HEGEWALD (2000) considers 11 taxa to be synonyms of this variety.

It has often not been possible to see the exact location of the poles, especially if the base of the spines is very thick and when the cell poles are more elongated. For this reason, many experts doubt the use of this character as a diagnosis. In the current study, eccentric localization of the spines was documented in all the specimens observed.

Desmodesmus opoliensis (P. Richter) Hegewald var. *carinatus* (Lemmermann) Hegewald (Fig. 89-91)

Algological Studies 96: 15. 2000.

Basionimo: *Scenedesmus opoliensis* var. *carinatus* Lemmermann, Forschungsberichte aus der Biologischen Station zu Plon 7, 113. 1899.

Synonym: *Scenedesmus carinatus* (Lemmermann) Chodat

Cenobium formed by 4 cells, cells arranged linearly, elliptical to fusiform, outer cells with 1 long spine at the poles, acute poles almost truncated, inner cells with rounded poles, ribs fragmented and sometimes with small denticles. Cell length 7.9-15.45 pm, diameter 2.0-4.75 pm, polar spines 9-27 pm. Parietal chloroplastid with a pyrenoid.

Geographical distribution in the state of Sao Paulo

IN LITERATURE: **Municipio de Sao Paulo** [SANT'ANNAETTL. 1989: 96, fig. 81, as *Scenedesmus carinatus* (Lemmermann) Chodat].

EXAMINED MATERIAL: **Municipality of Arujá**, Clube Fiscal do Brasil, lake, 01- V-1975, *L. Sormus* (SP130815). **Municipality of Guaratinguetá**, Clube dos 500, lake, 01-IV- 1966, *C.E.M. Bicudo* (SP96965). **Municipality of Juquiá**, BR-116, km 160, flooded, 01-III- 1973, *C.E.M. Bicudo & L. Sormus* (SP113672). **Municipality of Miguelópolis**, dam, 30-V- 2000, *C.E.M. Bicudo & D.C. Bicudo* (SP365690). **Municipality of Sao Paulo**, Horto Florestal, lake, 16-VII-1962, *C.E.M. Bicudo & R.M.T. Bicudo* (SP96850).

Comment

According to CHODAT (1926), *Desmodesmus opoliensis* (P. Richter) Hegewald var. *carinatus* (Lemmermann) Hegewald forms a cenobium very similar to that of *Desmodesmus opolienses* (P. Richter) Hegewald. Morphologically, both have the same general appearance, with four elliptical-fusiform cells, however, D. *opoliensis* (P. Richter) Hegewald var. *carinatus* (Lemmermann) Hegewald may have ribs that cut across the entire longitudinal axis of the cells or may appear fragmented, in addition to the presence of 1-3 denticles at the cell poles.

Desmodesmus opoliensis (P. Richter) Hegewald var. *carinatus* (Lemmermann) Hegewald is a variety that does not have many references in the specialized literature, nor is it common in the state of Sao Paulo.

Specimens of D. *opoliensis* (P. Richter) Hegewald var. *carinatus* (Lemmermann) Hegewald have been recorded in five localities in the state of Sao Paulo and can be considered easy to identify taxonomically. The representatives of this variety closely resemble those of the standard variety of the species, but those of var. *carinatus* (Lemmermann) Hegewald have fragmented ribs and two or three denticles at the cell poles.

The species showed morphological variation with regard to cell dimensions, the size of the spines and the presence or absence of small polar spines in the inner cells and ribs.

Desmodesmus opoliensis (P. Richter) Hegewald var. *mononensis* (R. Chodat) Hegewald (Fig. 92-94)

Algological Studies 96: 15. 2000.

Basionym: *Scenedesmus opoliensis* (P. Richter) Hegewald var. *mononensis* R. Chodat, Transactions ofthe Royal Society of South Africa 18: 31, fig. 6. 1929.

Cenobium formed by 4 cells; cells arranged linearly, elliptical to fusiform, outer cells with almost truncated acute poles, with 1 long spine at each pole, spines situated eccentrically at the cell poles, cells in contact almost throughout; outer cells may show concavity or 1 slight convexity in the median part of the outer margin. Cell length 7.0-14.9 pm, width 2.0-5.6 pm, polar spines 3-18 pm long; parietal chloroplastid, 1 pyrenoid.

Geographical distribution in the state of Sao Paulo

IN LITERATURE: **Municipality of Luiz Antônio** [PERES & SENNA 2000: 474, fig. 35, as *Scenedesmus protuberans* Fritsch]. **Municipality of Sao José dos Campos** [CARDOSO 1979: 90, fig. 111-117, 125, as *Scenedesmus opoliensis* P. Richter var. *opoliensis*]. **Municipio de Sao Paulo** (XAVERETAL. 1985: 183, fig. 79, as *Scenedesmus opoliensis* P. Richter). **Municipality of Teodoro Sampaio** (BICUDO *ET AL.* 1992: 301, fig. 44, as *Scenedesmus protuberans* Fritsch var. *protuberans*).

EXAMINED MATERIAL: **Municipality of Arujá**, Clube Fiscal do Brasil, lake, 01- V-1975, *L. Sormus* (SP130815). **Municipality of Miracatu**, faracatiá, Fazenda Pettena, lake, 01-III-1973, *C.E.M. Bicudo, C.R. Leite & L. Sormus* (SP113673). **Municipio de Pindamonhangaba,** BR-116, km 270-271, lagoon, 21-V-1966, *C.E.M. Bicudo* (SP96949). **Municipality of Santo André,** Estacao Biológica do Alto da Serra de Paranapiacaba, lake near the headquarters, 20-I-1966, *C.E.M. Bicudo* (SP96909). **Municipio de Sao Paulo**, Parque Estadual das Fontes do Ipiranga, Lago das Garbas, *M. Borduqui,* 12-VII-2006, 23°38'08"S and 23°40'18"S, 46°36'48"W and 46°38'00"W (SP399781). **Municipio de Sao Paulo**, Parque Estadual das Fontes do Ipiranga, Lago das Garbas, col. *M. Borduqui,* 17-I-2007, 23°38'08"S and 23°40'18"S, 46°36'48"W and 46°38'00"W (SP399782). **Municipality of Tambaú**, Clube de Tambaú, dam, 23-VI-1973, *D.M. Vital* (SP113574). **Municipality of Vargem Grande Paulista**, SP-270, km 42.4, on the right towards Cotia-Vargem Grande, chácara Ise, dammed stream, with masses of Cyanophyceae, on macrophytes, periphyton and phytoplankton, 18-II- 1992, *A.A.J. de Castro* (SP239138).

Comment

According to HEGEWALD & SILVA (1988), *Desmodesmus opoliensis* (P. Richter) Hegewald var. *mononensis* (R. Chodat) Hegewald differs from the typical variety of the species in that the cells touch each other for almost their entire length and the cenobia are organized more or less in lines, in addition to the presence of eccentric polar spines.

BICUDO *ET AL.* (1992) and PERES & SENNA (2000) identified as *Scenedesmus protuberans* Fritsch var. *protuberans* specimens that we are currently re-identifying as D. *opoliensis* (P. Richter) Hegewald var. *mononensis* (R. Chodat) Hegewald.

Specimens of D. *opoliensis* (P. Richter) Hegewald var. *mononensis* (R. Chodat) Hegewald have been documented in seven locations in the state of São Paulo, but they are not so easy to identify because they are very similar to the typical variety of the species. However, var. *mononensis* (R. Chodat) Hegewald has cells that touch each other laterally for most of its length. Finally, the variety shows great variation in the dimensions of the cell, the size of the spines and, sometimes, a slight convexity and/or concavity in the median part of the outer margin of the cell.

***Desmodesmusperforatus* (Lemmermann) Hegewald** (Fig. 95)

Algological Studies 96: 15. 2000.

Basiönimo: *Scenedesmus perforatus* Lemmermann, Zeitschrift für Fischerei und deren Hilfswissenschaften 11: 104. Fig. 3. 1903.

Cenobium flat, formed by 4-8 cells; cells arranged linearly, elliptical, poles rounded, outer cells with concave inner margin, convex outer, 1 spine at each pole; inner cells with concave margins, between them a lenticuliform spine; cell length 13.2-15.6, width 5.5-6.3 pm, polar spines 8-15.5 pm long; parietal chloroplastidium, 1 pyrenoid.

Geographical distribution in the State of Sao Paulo

IN LITERATURE: **Municipalities of Rio Claro** and **Sao Paulo** (SANT'ANNA 1984: 233, fig. 155, as *Scenedesmusperforatus* Lemmermann).

MATERIAL EXAMINED: only material from the literature.

Comment

According to SANT'ANNA (1984), *Desmodesmus perforatus* (Lemmermann) Hegewald has a lenticuliform spine between the cells, which is unique to the species. In addition, the outer cells have a concave inner margin and a convex outer margin.

***Desmodesmus pleiomorphus* (Hindák) Hegewald** (Fig. 96-97)

Algolgical Studies, 16. 2000.

Basionimo: *Scenedesmuspleiomorphus* Hindák, Archiv fur Hydrobiologie, Suppl. 79: 482, fig. 19i. 1988.

Flat cenobium formed by 4 cells; cells arranged linearly, elliptical to cylindrical, poles rounded; outer cells with 1 thorn at each pole, in addition to this 1(2) thorns at the margin of the outer cell; inner cells with 1 long thorn at each pole; cell length 9.7-12.3 pm, width 2.6-3.4 pm, polar spines 5-12 pm long;

parietal chloroplastid, 1 pyrenoid.

Geographical distribution in the state of Sao Paulo

IN THE LITERATURE: nothing. First mention of the occurrence of the species in the state of Sao Paulo.

EXAMINED MATERIAL: **Municipality of Fartura**, SP-287, direction Fartura- Taquarituba, km 3, stream, 26-VII-2000, *S.M.M. Faustino & S.P. Schetty* (SP365696). **Municipality of Guaratingueta**, Clube dos 500, lake, 01-IV-1966, *C.E.M. Bicudo* (SP96965). **Municipality of Pindamonhangaba**, BR-116, km 270-271, lake, 21-V-1966, *C.E.M. Bicudo* (SP96949). **Municipality of Santo André**, Estacao Biologica do Alto da Serra de Paranapiacaba, lake near the headquarters, 20-I-1966, *C.E.M. Bicudo* (SP96909). **Municipality of Sao Paulo**, Parque Estadual das Fontes do Ipiranga, Lago do IAG, 03-VIII-1998,*I.S. Vercellino,* 23°38'08"S and 23°40'18"S, 46°36'48"W and 46°38'00"W (SP399779); 06-III-1999, *I.S. Vercellino,* 23°38'08"S and 23°40'18"S, 46°36'48"W and 46°38'00"W (SP399780); Lago das Garbas, 12-VII-2006, *M. Borduqui,* 23°38'08"S and 23°40'18"S, 46°36'48"W and 46°38'00"W (SP399781); 17-I-2007,*M Borduqui, 23°38'*08"S and 23°40'18"S, 46°36'48"W and 46°38'00"W (SP399782).

Comment

Desmodesmus pleiomorphus (Hindák) Hegewald morphologically resembles Desmodesmus *spinosus* (Chodat) Hegewald and *Desmodesmus abundans* (Kirchner) Chodat, however, it differs from both by having much larger spines on the inner cells, which is a strong characteristic of the species.

HINDÁK (1990) observed high morphological variability in material from this species in cultivation (Hindák strain n^0 1982/24 isolated from samples collected in Austria), detecting cenobia with spines, ribs and granules. The author also observed that in cenobia with short spines, ribs were formed and, on the contrary, in cenobia with long spines, ribs were absent.

There is little information on D. *pleiomorphus* (Hindák) Hegewald in both Brazilian and world literature due to its uncommon occurrence in the materials collected so far. The species is universally considered rare. In addition, in all the material examined, only specimens with long spines and without ribs were noted. Specimens have currently been collected in five localities in the state of Sao Paulo, but always in the form of a few popular individuals.

Desmodesmuspleiomorphus (Hindák) Hegewald showed no significant morphological variation in the waters of the state of São Paulo, except for the fact that it presented two spines on the outer margin of the cells and always kept the character of long spines on the cell poles of the inner cells in the popular ones analyzed.

Desmodesmus protuberans **(Fritsch & Rich) Hegewald** (Fig. 98)

Algological Studies 96: 16. 2000.

Basionimo: *Scenedesmusprotuberans* **Fritsch & Rich**, Transactions of the Royal Society of South Africa 18: 31, fig. 6. 1929.

Cenobia flat, formed by 4 cells; cells arranged linearly, elliptic-fusiform to fusiform, poles prominent, rounded, outer cells with convex margins in the median part, inner cells without spines and smaller than the outer ones; cell length 10.0-39.0 pm, width 3.5-10.0 pm; spines 8-20 pm long; parietal chloroplastidium, without pyrenoid.

Geographical distribution in the state of Sao Paulo

IN LITERATURE: **Municipality of Santo André** and **Sao Bernardo do Campo** (SANT'ANNA 1984: 234, fig. 156, as *Scenedesmusprotuberans* Fritsch var. *protuberans* f. protuberans*).* **Municipio de Sao Paulo** (SANT'ANNA *ET AL.* 1989: 97, fig. 91, as *Scenedesmusprotuberans* Fritsch).

MATERIAL EXAMINED: only material from the literature.

Comment

According to COMAS (1996), this species is very similar in morphology to *Desmodesmus opoliensis* (P. Richter) Hegewald, from which it differs fundamentally by its more prominent, more or less capitate poles and by the spines located approximately in the center of the cell poles. SANT'ANNA (1984) points to the morphology of the cell poles as the only distinguishing feature between *D. opoliensis (*P. Richter) Hegewald and *D. protuberans* (Fritsch & Rich) Hegewald, stating that in *D. opoliensis* (P. Richter) Hegewald they are eccentric. This species can be confused with *D. opoliensis,* from which it differs fundamentally by the more projecting cell ends and the insertion of the spines at the cell poles, which do not make the pole angles, but are more or less in the middle.

Desmodesmuspseudodenticulatus **(Hegewald) Hegewald** (Fig. 99-100)

Algological Studies 96: 16. 2000.

Basiönimo: *Scenedesmus pseudodenticulatus* **Hegewald** in Hegewald & Schnepf, Algological Studies

20: 312, 315, fig. 6-17. 1978.

Cenobium flat, formed by 4 cells; cells arranged linearly, oblong, poles rounded or slightly truncated, outer cells with small, independent spines, slightly distant from each other at the outer margin of the cell, 1-2 spines at the poles, inner cells with 1-2 denticles at the poles; cell length 12.0-13.8 pm, width 3.7-5.55 pm; parietal chloroplastidium, 1 pyrenoid.

Geographical distribution in the State of Sao Paulo

IN LITERATURE: **Municipio de Sao Paulo** [FERRAGUT *etAL.* 2005: 154, fig. 83, as *Desmodesmuspseudodenticulatus* (Hegewald) Hegewald].

EXAMINED MATERIAL: **Municipality of Campos de Jordao**, Alagoinha neighborhood, lake, 11-III-1973, *M.M. Sakane* (SP130445). **Municipality of Rio Claro**, SP-310, km 156, flooded, 10-V-1973, *C.E.M. Bicudo & P.A.C. Senna* (SP104728). **Municipio de Sao Paulo**, Parque Estadual das Fontes do Ipiranga, Lago do IAG, 03-VIII-1998, *7.S. Vercellino,* 23°38'08"S and 23°40'18"S, 46°36'48"W and 46°38'00"W (SP399779); Lago das Ninféias, 03- VIII-2007, *T.R. Santos,* 23°38'08"S and 23°40'18"S, 46°36'48"W and 46°38'00"W (SP399783). **Municipio *de* Sao Pedro**, SP-304, km 127, lake of Restaurante do Lago, with *Nymphaea elegans* and *Salvinia,* periphyton, 20-III-1990, *A.A.J. de Castro & C.E.M. Bicudo* (SP255724).

Comment

According to COMAS (1996), *Desmodesmus pseudodenticulatus* (Hegewald) Hegewald falls into a group of species based on the ultrastructure of the cell wall. It has a granular cell wall, with tiny spines on the margin of the outer cell. Although FERRAGUT *ET al.* (2005) state that they observed a granular cell wall in the specimens observed, no ornamented cell wall was found in any other specimen collected in the state of Sao Paulo.

HEGEWALD & SILVA (1988) stated that cenobia with eight cells or with a series of denticles in the median cells have never been observed. We agree with this author, as these characteristics were also never observed in the samples from the state of Sao Paulo. The specimens examined by FERRAGUT *ETAL.* (2005) are very similar to these in terms of the outer margin of the outer cells with small, independent spines, slightly distant from each other, as well as the presence of one or two spines at the poles and the inner cells with one or two denticles at the poles.

Specimens of D. *pseudodenticulatus* (Hegewald) Hegewald were collected from four sites in the state of Sao Paulo. The species was not easy to identify, mainly because the shape of the cells, spines and denticles are similar to those of *D. lunatus* (West & West) Hegewald and *D. spinulatus* (Biswas) Hegewald. However, it differs from these two species in having oblong cells with truncated poles.

***Desmodesmus serratus* (Corda) An, Friedl & Hegewald** (Fig. 101-104)

Algological Studies 96: 17. 2000.

Basionym: *Arthrodesmus serratus* Corda, Almanach de Carlsbad 9: 244, pl. VI, fig. 35. 1839.

Synonym: *Scenedesmus serratus* (Corda) Bohlin

Cenobium formed by 4 cells; cells arranged linearly, elliptical to elliptical-fusiform, poles rounded, outer cells with small spines on the lateral margins, but no surrounding membrane; inner cells with 1 small spine at the poles, longitudinal ribs in the median part of the cell, ribs formed by joining very small spines; cell length 7.3-21.3 pm, width 2.1-7.9 pm, polar spines 2-3 pm long; parietal chloroplastid, 1 pyrenoid.

Geographical distribution in the state of Sao Paulo

IN LITERATURE: **Municipio de Sao Paulo** (PEFI), [SANT'ANNA *ET AL.* 1989: 97, fig. 95, as *Scenedesmus serratus* (Corda) Bohlin].

EXAMINED MATERIAL: **Municipality of Juquiá**, BR-116, km 160, flooded, 01-III- 1973, *C.E.M. Bicudo & L. Sormus* (SP113672). **Municipio de Macedonia**, acude, rodovia Alberto Faria, direction Mia Estrela-Macedonia, 2 km before the entrance to Macedonia, 25-IV- 2001, *C.E.M. Bicudo, D.L. Costa & S.M.M. Faustino,* GPS 20° 08'19,5"S, 50°11'56,4"W, Conductivity 70 pS.cm⁻¹ , pH 6,6, (SP355366). **Municipality of Pirassununga**, SP-225, km 23, Cascalho neighborhood, lake, ?-XII-1973, *P.A.C. Senna* (SP123900). **Municipality of Santo André**, Estacao Biológica do Alto da Serra de Paranapiacaba, stream near the headquarters, 20-I-1966, *C.E.M. Bicudo* (SP96922). **Municipio de Sao Carlos**, SP-310, km 222, Aldeia Conde do Pinhal, stream, 10-V-1973, *C.E.M. Bicudo &L. Sormus* (SP104723).

Comment

CHODAT (1926) observed cenobia with four cells as in the current specimens, while SANT'*ANNA* ET AL. (1989) identified cenobia with two cells.

Desmodesmus serratus (Corda) An *et al.* is characterized by the presence of a row of denticles on each lateral margin, running longitudinally from one to the other pole; however, this character is very variable and sometimes only one denticle can be seen at each pole under light microscopy. In this case, D. *serratus* (Corda) An *et al.* can be confused with other species such as *Scenedesmus brevispina* (G.M. Smith) Chodat [= *D. denticulatus* (Lagerheim) An *et al.* var. *linearis* (Hansgirg) Hegewald, according to HEGEWALD (2000)].

Desmodesmus serratus (Corda) An *et al.* can also be confused with *Desmodesmus pseudodenticulatus* (Hegewald) Hegewald, however, the latter species does not have ribs cutting through the entire cell wall or fragmented. However, the two species have independent spines along the entire length of the outer margin of the cell, as the cell shape of the two species is very different: D. *serratus* (Corda) An *et al.* has elliptical to elliptical-fusiform cells with rounded cell poles and *D. pseudodenticulatus* (Hegewald) Hegewald has oblong cells with rounded or sometimes truncated cell poles. In addition to the similarity with D. *denticulatus* (Lagerheim) An *et al.* var. *linearis* (Hansgirg) Hegewald and D. *pseudodenticulatus* (Hegewald) Hegewald, the species in question also bears a close resemblance to D. *brasiliensis* (Bohlin) Hegewald, however, D. *serratus* (Corda) An *et al.* has spines close to the outer margin of the cell, but independent. It also differs by the denticle-shaped ribs that cut across the entire length of the cell, a feature that does not occur in D. *brasiliensis* (Bohlin) Hegewald; and by the shape of the cell, which is elliptical-fusiform and in D. *brasiliensis* oblong.

Desmodesmus serratus (Corda) An *et al.* occurs in five localities in the state of Sao Paulo and the current specimens have always been found in large populations. However, it is a difficult species to identify, as it is very similar to several others in the genus. Great care must be taken when identifying this species, as it has shown significant morphological variation in terms of cell shape, cell wall ornamentation, the number of spines at the cell poles and the presence or absence of ribs. We have even observed individuals without denticles on the outer margin of the cell; for this reason, we once again state that D. *serratus* (Corda) An *et al.* is a highly variable species that is difficult to identify.

Desmodesmus spinosus (R. Chodat) Hegewald (Fig. 105-106)
Algological Studies 96: 17. 2000.
Basionimo: *Scenedesmus spinosus* R. Chodat, Monographies d'algues en culture pure. 74, fig. 70-74. 1913.

Flat cenobium formed by 4 cells; cells arranged linearly, elliptical to cylindrical, poles rounded, outer cells with 1-2 spines on each pole, 1 spine on the margin of the outer cell. Cell length 9.7-12.3 pm, width 2.6-3.4 pm, polar spines 5-12 pm long; parietal chloroplastid, 1 pyrenoid.

Geographical distribution in the State of Sao Paulo

IN LITERATURE: **Municipalities of Arujá, Atibaia, Rio Claro, Salesópolis, São Paulo, Sumaré and Tambaú** [SANT'ANNA 1984: 249, fig. 165-166, as *Scenedesmus spinosus* Chodat]. **Municipality of Ribeirao Preto** [SILVA 1999: 291, fig. 67, as *Scenedesmus spinosus* Chodat]. **Municipality of Sao José dos Campos** [CARDOSO 1979: 102, fig. 130-132, as *Scenedesmus spinosus* R. Chodat var. *spinosus].* **Municipality of Sao Paulo** [MOURA 1996: 49, fig. 19, as *Scenedesmus spinosus* Chodat; TUCCI 2002: 243, fig. 49, as *Scenedesmus spinosus* Chodat, FERRAGUT *ET AL.* 2005: 153, fig. 82, as *Desmodesmus polyspinosus* (Hortobágyi) Hegewald, FERRAGUT *ETAL.* 2005: 154, fig. 84-85, *Desmodesmus spinosus* (Chodat) Hegewald var. *spinosus', TUCCI ETAL.* 2006: 165, fig. 52, as *Scenedesmus spinosus* Chodat].

MATERIAL EXAMINED: only material from the literature.

Comment

Desmodesmus spinosus (R. Chodat) Hegewald is very similar to *Desmodesmus abundans* (Kirchner) Hegewald, to the point that some authors (e.g. PHILIPOSE 1967) consider the first species to be a synonym of the second. SMITH (1916b) did not consider S. *spinosus* (R. Chodat) Hegewald a synonym of .S' *nanus* Chodat, but SANT'ANNA (1984) did not after considering all the controversy in the literature and making a comparative study of the characters used to delimit the two species.

What differentiates D. *spinosus* (R. Chodat) Hegewald from D. *abundans* (Kirchner) Hegewald is the presence of two spines of unequal size on the outer cells. Sometimes, these spines can occur alone and, although sporadically, also in numbers of three. In D. *spinosus* (R. Chodat) Hegewald, the marginal *spines* are practically the same size and appear in greater numbers. In addition, there may be spines throughout the cell wall of all the cells in the cenobium. Other distinguishing features between these two species are the shape of the cenobium cells, which in D. *spinosus* (R. Chodat) Hegewald are elliptical-cylindrical and in D. *abundans* (Kirchner) Hegewald oblong, and the presence of a convexity on the outer margin of the extreme cells of the cenobium. What makes it difficult to separate D. *spinosus* (R. Chodat) Hegewald from D. *abundans* (Kirchner) Hegewald is that the spines can sometimes not be seen on the cell wall, as happened with the present specimens recorded in the literature from the state of Sao Paulo, making it difficult to identify D. *spinosus* (R. Chodat) Hegewald.

Desmodesmus spinosus (R. Chodat) Hegewald is not only easily confused with *D. abundans* (Kirchner) Hegewald, it is also commonly confused with *D. subspicatus* (R. Chodat) Hegewald & Schmidt. It differs, however, in that *D. spinosus* (Kirchner) Hegewald has lateral spines and shows a tendency to disaggregate the cenobia into solitary cells. *Desmodesmus subspicatus* (R. Chodat) Hegewald & Schmidt lacks ribs and has a greater number of lateral spines (2-5) in the marginal cells, reminiscent of the current specimens from the state of Sao Paulo (fig. 56). However, *Desmodesmus* sp. 1 has thickening at the cell poles, a feature that does not occur in D. *subspicatus* (R. Chodat) Hegewald & Schmidt.

FERRAGUT *ET AL.* (2005) identified certain specimens with *Desmodesmus polyspinosus* (Hortobágyi) Hegewald, but on evaluating the illustration we concluded that it is D. *spinosus* (R. Chodat) Hegewald. TUCCI *ET AL.* (2006) identified *Scenedesmus spinosus* Chodat; however, the specimens they examined differed from those found in the state of São Paulo in that they had three spines on the margins of the outer cells, which is not very common in the specimens currently examined.

Desmodesmus spinulatus (**Biswas) Hegewald** (Fig. 107-109)

Algological Studies 96: 17. 2000.

Basionym: *Scenedesmus spinulatus* **Biswas**, Hedwigia 74: 20. 1934.

Cenobium flat, formed by 4 cells; cells arranged linearly or alternately, oval-fusiform, poles truncated to rounded, outer cells with 1-3 small spines at the poles, 1 spine usually projected perpendicularly to the cenobium, can also occur in the inner cells, however, it is more common in the outer cells, margin of outer cells with independent small spines, inner cells with 1-2 small spines at the poles; smooth cell wall, apart from ribs which may or may not be present; cell length 10.0-16.2 pm, width 3.0-3.75 pm, polar spines 1-4 pm long; parietal chloroplastidium, 1 pyrenoid.

Geographical distribution in the state of Sao Paulo

IN LITERATURE: Municipality of Luiz Antonio (PERES & SENNA 2000: 474, fig. 27, as *Scenedesmus denticulatus* Lagerheim).

EXAMINED MATERIAL: **Municipality of Campos de Jordao**, Alagoinha neighborhood, lake, 11-III-1973, *M.M. Sakane* (SP130445). **Municipality of Itu**, SP-280, km 77, lake, 11-V-1977, *C.R. Leite* (SP139733). **Municipality of Rio Claro**, Horto Florestal, lake, 31-I-1975, *O.A. da Silva* (SP123867). **Municipality of Sao Paulo**, Parque Estadual das Fontes do Ipiranga, Lago das Garças, 17-I-2007, *M. Borduqui,* 23°38'08"S and 23°40'18"S, 46°36'48"W and 46°38'00"W (SP399782).

Comment

Desmodesmus spinulatus (Biswas) Hegewald can be confused with *Desmodesmus lunatus* (West & West) Hegewald, however, what differentiates them are the outer cells of the cenobium of the first species with one to three small spines on the poles. A thorn usually projected perpendicularly to the cenobium can also occur on the inner cells, but is more common on the outer cells. The margins of the outer cells of the cenobium of *D. spinulatus* (Biswas) Hegewald may have independent small spines and the inner cells may have one or two small spines at the poles. The cell wall can be smooth or granular. *Desmodesmus spinulatus* (Biswas) Hegewald can also be confused with *D. brasiliensis* (Bohlin) Hegewald, however, the latter species does not have the perpendicular spines characteristic of D. *spinulatus* (Biswas) Hegewald.

This species is distributed in tropical and subtropical countries and is readily recognized by its relatively oval-fusiform cells, aligned or alternating with more or less vigorous denticles at the cell poles. Finally, the marginal cells have a characteristic orientation, i.e. one axis is oriented perpendicular to the cenobium while the other forms an almost right angle to the first. The marginal cells may have a row of delicate teeth.

PERES & SENNA (2000) identified material collected in the municipality of Luiz Antonio as *Scenedesmus denticulatus* Lagerheim, but on re-examining this material we concluded that it was a representative of *Desmodesmus spinulatus* (Biswas) Hegewald. The fact is that the two species are really similar, as they have cenobiums with alternating cells, but what differentiates them is the shape of the cell, which in *D. spinulatus* (Biswas) Hegewald is elliptical-fusiform, with truncated poles.

According to COMAS *ETAL.* (2007), the species currently included in D. *spinulatus* (Biswas) Hegewald have a very high morphological similarity to *Scenedesmus polydenticulatus* Hortobágyi due to the vigorous and laterally free spines present on the margins of the outer cells of the cenobium.

Desmodesmus spinulatus (Biswas) Hegewald was collected in samples from four localities in the state of Sao Paulo. However, the species showed great morphological variability with regard to the shape of the cells (some individuals had an outer cell with a concavity in the center of the margin and an alternating cenobium, very similar to D. *denticulatus),* the number of spines on the cell poles and the ornamentation of the cell wall. Also, the species was not easy to identify; on the contrary, it is a species that requires a lot of caution when identifying it, mainly because it is morphologically very similar to other species in the genus.

Desmodesmus subspicatus **(R. Chodat) Hegewald & Schmidt** (Fig. 110-111)
Algological Studies 96: 17. 2000.
Basionym: *Scenedesmus subspicatus* **Chodat**, Zeitschrift für Hydrologie 3: 222, fig. 128.
1926.

Cenobium flat, formed by 2 cells; cells arranged linearly, elliptical, poles rounded, outer cells with 1 long spine at each pole, margins composed of spines of different sizes, unequal to each other, independent, well apart from each other, 1 spine at the poles just below the longest spines; cell length 8.3-9.8 pm, width 3.45-4.35 pm; parietal chloroplastid, 1 pyrenoid.

Geographical distribution in the state of Sao Paulo

IN LITERATURE: **Municipality of Sao José dos Campos** (CARDOSO 1979: 100, fig. 128-129, as *Scenedesmus spicatus* West & West var. spicatus; CARDOSO 1979: 104, fig. 133, as *Scenedesmus spinosus* Chodat var. *bicaudatus* Hortobágyi).

EXAMINED MATERIAL: **Municipio de Sao Paulo**, Horto Florestal, lake, 16-VII- 1962, *C.E.M. Bicudo &R.M.T Bicudo* (SP96850).

Comment

Desmodesmus subspicatus (R. Chodat) Hegewald & Schmidt is very similar to *Desmodesmus spinosus* (R. Chodat) Hegewald, however, the former does not have spines all over the cell wall like D. *spinosus* (R. Chodat) Hegewald and the cell shape is elliptical.

Desmodesmus subspicatus (R. Chodat) Hegewald & Schmidt includes, according to HEGEWALD (2000), 31 taxa as synonyms and three of them were identified from material in the literature from the state of Sao Paulo because of their oblong, somewhat elongated cells with rounded poles that generally have two or three longitudinally arranged lateral spines.

CARDOSO (1979) identified material of *Scenedesmus spicatus* West & West var. *spicatus* and *Scenedesmus spinosus* Chodat var. *bicaudatus* Hortobágyi, but on re-examining the illustrations of this material we concluded that it was D. *subspicatus* (R. Chodat) Hegewald & Schmidt.

We concluded that in the current samples from the state of Sao Paulo, we observed populations that corresponded, for the most part, to the material of D. *subspicatus* (R. Chodat) Hegewald & Schmidt (fig. 53a-b), alongside others that corresponded to *Scenedesmus gutwinskii* Chodat var. *bekesensis* Uherkovich (fig. 76a-b). The latter should, however, be included in D. *subspicatus* (R. Chodat) Hegewald & Schmidt as an independent variety to be renamed, since *bekesensis* is not a valid name because it has no Latin diagnosis or designated nomenclatural type (see art. 36 and 37 of the International Code of Botanical Nomenclature).

Desmodesmus subspicatus (R. Chodat) Hegewald & Schmidt was only found in the municipality of Sao Paulo, in a lake in the Horto Florestal. This is because it occurred in a single municipality, with a population consisting of only a few individuals. *Desmodesmus subspicatus* (R. Chodat) Hegewald & Schmidt was not considered easy to identify due to its great morphological similarity to D. *spinosus* (R. Chodat) Hegewald. With regard to morphological variation, D. *subspicatus* (R. Chodat) Hegewald & Schmidt showed changes in the number of spines on the margin of the outer cell and in the shape of the cell, which is generally elliptical, but in some cases was oblong, with slightly acute poles; however, the polar spines were always present.

***Desmodesmus* sp. 1** (Fig. 112)

Cenobium flat, formed by 4 cells; cells arranged linearly, elliptical, poles rounded, outer cells with 1 long thorn on each pole, another thorn in the central region of the outer margin, 1-2 small thorns on the lateral poles of all the cells of the cenobium, thickening on the cell poles of all the cells; cell length 7.3-9.0 pm, width 3.2-3.7 pm; parietal chloroplast, 1 pyrenoid.

Geographical distribution in the State of Sao Paulo

IN THE LITERATURE: nothing. First mention of the species in Brazil.

EXAMINED MATERIAL: **Municipality of Batatais**, SP-330, km 355.5, right-hand side, direction Batatais-Franca, in front of the 'Aparecida' seedling nursery, dam, with *Hydrocotile* and *Myriophyllum,* periphyton, *A.A.J. de Castro,* 16-XI-1991 (SP255761). **Municipality of Campos de Jordáo**, Alagoinha neighborhood, lake, 11-III-1973, MM *Sakane* (SP130445).

Comment

The presence of *Desmodesmus* sp. 1 was recorded in two municipalities in the state of São Paulo: Batatais and Campos do Jordao. It was not considered an easily identifiable species and, despite having characteristics very similar to those of other species such as D. *abundans* (Kirchner) Hegewald and D. *spinosus* (R. Chodat) Hegewald, we concluded that it was a new species due to the presence of lateral polar spines and a thickening of the cell wall in all the cells of the cenobium. Few individuals representing this species were found in the two small populations, but the constancy of the above characteristics was absolute. It is absolutely necessary, however, to study more specimens of this material in order to confirm its identification as a taxonomic novelty.

***Desmodesmus* sp. 2** (Fig. 113)

Cenobium flat, formed by 4 cells; cells linearly arranged, elliptical, poles rounded, outer cells with granules on outer margin, 1-2 spines on lateral poles of all cenobium cells; cell length 6.2-7.8 pm, width 2.52.9 pm; parietal chloroplastid, 1 pyrenoid.

Geographical distribution in the State of Sao Paulo

IN THE LITERATURE: nothing. First mention of the species in Brazil.

EXAMINED MATERIAL: **Municipality of Pindamonhangaba**, Sao Joao farm, Sao Joao lagoon, 21-V-1966, *C.E.M. Bicudo* (SP96950).

Comment

Desmodesmus sp. 2 was collected in only one municipality in the state of São Paulo, Pindamonhangaba, and the population sampled consisted of only a few individuals. *Desmodesmus* sp. 2 was not considered an easily identifiable species because it has characteristics that are very similar to those of other species, such as *D. lunatus* (West & West) Hegewald. The latter species also has small teeth on the margin of the outer cells and these teeth can in fact be considered warts or granules. However, what differentiates *Desmodesmus* sp. 2 from *D. lunatus* (West & West) Hegewald is the fact that the former has a granular outer margin, with one or two denticles at the poles of all the cenobium cells. *Desmodesmus sp.* 2 can be described as a new species, but it is essential to examine a greater number of individuals in order to know the constancy of the morphological characteristics indicated above as diagnostic of the population studied.

***Desmodesmus* sp. 3** (Fig. 114-115)

Cenobium curved, formed by 4 cells; cells arranged alternately, elliptical, poles rounded; outer cells with 1 long thorn on each pole, another tiny "tooth" shaped thorn lying on the lateral poles; cell length 5.6-8.2 pm, width 1.2-3.4 pm; parietal chloroplastidium, 1 pyrenoid.

Geographical distribution in the State of Sao Paulo

IN THE LITERATURE: nothing. First mention of the species in Brazil.

EXAMINED MATERIAL: **Municipality of Arujá**, Clube Fiscal do Brasil, lake, 01- V-1975, *L. Sormus* (SP130815).

Comment

Desmodesmus sp. 3 was collected in a single municipality in Sao Paulo, Arujá. A population consisting of a few specimens was sampled, which were not easy to identify taxonomically due to the fact that their characters were very similar to those of other species. What differentiated *Desmodesmus* sp. 3 from the other species in the genus were the small spines or "tooth" shaped denticles located at the cell poles which can occur in all the cells of the cenobium or in only a few, but which were always present in the specimens in the population. Another fact that caught our attention was the great variation in the presence of a thorn on one of the cell poles, diagonally opposite if we consider the cenobium, which can confuse the identification of *Desmodesmus* sp. 3 with D. *armatus* (R. Chodat) Hegewald var. *bicaudatus* (Guglielmetti) Hegewald or D. *intermedius* (R. Chodat) Hegewald var. *acutispinus* (Roll) Hegewald, which are the two other species that have this bicaudate feature. In this case, the bicaudate character in *Desmodesmus* sp. 3 would be a morphological variation of the species. Hence the imperative need to study the popular ones for the taxonomic identification of species, varieties and taxonomic forms and, sometimes, even genera of microalgae. Hence the need to analyze more specimens of this type in order to consider them as representatives of a species new to science.

Pseudodidymocystis Hegewald & Deason 1989

The cenobium is made up of just two cells which are arranged next to each other along their longest axes. A relatively abundant mucilaginous matrix surrounds the cenobium. The cells are ellipsoid to semicircular. The cell wall is ornamented in the form of granular inorganic material or warts. Parietal chloroplastidia with or without pyrenoids (COMAS 1996).

Key to identifying the species studied:

1. Smooth cell wall...*Thin P.*
1. Ornate cell wall .... *Planktonic P.*

Pseudodidymocystis fina (Komárek) Hegewald & Deason (Fig. 116)
Algological Studies 55: 127. 1989.
Basionimo: _Didymocystis fina_ Komárek, Preslia 47: 276, fig. 3. 1975.

Cenobium flat, formed by 2 cells; cells arranged linearly, oblong, outer margins convex or almost straight; cell wall smooth; cell length 5.210.2 pm, width 2.3-4.1 pm; chloroplastidium parietal, without pyrenoid.

Geographical distribution in the State of Sao Paulo

IN LITERATURE: **Municipality of Sao Paulo** (Tuca 2002: 243, fig. 41; Ferragut *et al.* 2005: 154, fig. 87, as *Didymocystisfina* Komárek; TUCCI ET*AL.* 2006: 163, fig. 46).

EXAMINED MATERIAL: **Municipality of Ara^atuba**, Marechal Rondon highway, unspecified location, with Cyperaceae, grasses and *Myriophyllum,* 15-I-1992, *L.H.Z. Branco* (SP239239). **Municipio de Guapiara**, SP-250, km 284, rio Sao José, periphyton, 27- III-2001, *C.E.M. Bicudo, L.A. Carneiro & S.MM. Faustino,* 24°19'12.0"S, 48°37'1.7"W, conductivity 30 pS cm^{-1} , pH 6.9 (SP355391). **Municipality of Juquiá**, BR-116, km 160, flooded, 01-III-1973, *C.E.M. Bicudo & L. Sormus* (SP113672). **Municipio de Mogi das Cruzes**, SP-88, km 74-75, lagoon, 18-VI-1973, *C.E.M. Bicudo, C.R. Leite & L. Sormus* (SP113662). **Municipio de Sao Paulo**, Parque Estadual das Fontes do Ipiranga, Lago do IAG, 03-VIII-1998, I.*S. Vercellino,* 23°38'08"S and 23°40'18"S, 46°36'48'W and 46°38'00'W (SP399779); Lago das Ninféias, 03-VIII-2007, *T.R. Santos,* 23°38'08"S and 23°40'18"S, 46°36'48"W and 46°38'00"W (SP399783).

Comment

KOMÁREK (1973) considered the generic diagnostic character of *Didymocytis to be the formation of* two-celled cenobia, so that one mother cell would form four autospores and these autospores would form two cenobia, each with two cells. *Scenedesmus* can form two-cell cenobia, however, if they formed four autospores, these would be united to form a single cenobium. For KOMÁREK (1973), the pyrenoid was of no importance in separating the two genera. Consequently, KOMÁREK (1973) designated *Didymocystisplanctonica* Korsikov the type species of the genus.

The principle of priority has a somewhat problematic and questionable application in this case. It is accepted that FOTT (1973) has priority, because both articles were published in 1973, in Preslia volume[0] 45, but Fott's article was on pages 1-10 and Komárek's on pages 311-314. We don't know for sure, but it seems that the editor received Fott's work first. This fact is of no importance, however, because what counts is the date the article was published. If the date of publication is exactly the same, as in this case, any objective criterion can prevail in the choice of nomenclatural type and the pagination reflects, to a certain extent, an anticipation: that of the acceptance of the work for publication. In conclusion, the type species of *Didymocystis* should be *D. inermis* Korsikov and not D. *planctonica* Korsikov.

The genus *Didymocystis* is characterized by the formation of two-celled cenobia. It was included in *Scenedesmus* by FOTT (1973) and HEGEWALD (1978), but KOMÁREK (1973) and Komárek & Fott (1983) preferred to keep it independent. HEGEWALD (1988) showed that the type species of *Didymocystis* is different from all the others included in the genus because it has a cell wall with cellulose fibers and does not contain sporopollenin. The *Scenedesmus* species have a cell wall with sporopollenin. Studies of the cell wall in cultures of *Didymocystis fina* Komárek revealed special structures in the cell wall.

According to HEGEWALD & DEASON (1989), *Didymocystis fina* Komárek was described without pyrenoid or mucilage, however, there is a layer of mucilage whose thickness is less than 1 pm and, perhaps for this reason, it is easily overlooked. The pyrenoid is also not always visible, according to the same authors. KOMÁREK (1975, 1983) and COMAS (1991) did not observe pyrenoids in the materials they studied. HEYNIG & DRIENITZ (1987), HEYNIG (1989) and HEGEWALD & DEASON (1989) observed pyrenoids in material collected in Europe and considered identical to *Pseudodidymocystisfina* (Komárek) Hegewald & Deason.

Hindák (1984a) transferred *Pseudodidymocystis fina* (Komárek) Hegewald & Deason to the genus *Choricystis* based on optical microscopy studies of material from Cuba identified as *Didymocystis fina* Komárek (see HINDÁK 1984a: pl. 71, fig. 6). However, COMAS (1996) did not have the type material of *Pseudodidymocystis fina* (Komárek) Hegewald & Deason available for study. Instead, he studied popular specimens that fully corresponded to the original diagnosis of the species. COMAS (1996) concluded that the material evaluated is a cenobium-forming alga, which cannot be identified with *Choricystis*.

According to TUCCI *ET AL.* (2006), the specimens from the Parque Estadual das Fontes do Ipiranga, in the municipality of São Paulo, have elongated cells arranged parallel to each other in relation to their longitudinal axis. The cells have a smooth cell wall. We agree with TUCCI ET al. (2006), as all the specimens currently studied have this characteristic, except for their size, since the current specimens have been consistently slightly larger.

Representatives of this species were found in five localities in the state of Sao Paulo; however, the species is not considered easy to identify due to the fact that it is very similar to others that have two cells in the cenobium and no spines. Intraspecific morphological variation was only noted in relation to cell dimensions, but these were never significant enough to influence its taxonomic identification.

Pseudodidymocystis planctónica (Korsikov) Hegewald & Deason (Fig. 117)
Algological Studies 55: 127. 1989.
Basionimo: *Didymocystis planctónica* Korsikov, Viznacnik prisnovodnich vodorostej Ukrainskoj RSR 5: 396, fig. 399.1953.

Cenobium flat, formed by 2 cells; cells arranged linearly, oblong, outer margins convex or almost straight; cell wall may show granulacbes; cell length 8.0-13.4, width 3.8-5.5 qm; parietal chloroplastid, 1 pyrenoid.

Geographical distribution in the State of Sao Paulo

IN LITERATURE: **Municipality of Sao Paulo** (MOURA 1996: 46, fig. 6, as *Didymocystisplanctónica* Korsikov; TUCCI2002: 243, fig. 40; TUCCIE7'. IL. 2006: fig. 47).

MATERIAL EXAMINED: **Municipality of Pitangueiras**, SP-322, km 368, acude, 16-VII-2000, *C.EM. Bicudo, SMM. Faustino & L.L. Morandi*, 20°59'30.5"S, 48°14'01.1"W, conductivity 40 qS cm^{-1}, pH 6.5 (SP355382). **Municipio de Itaberá**, SP- 249, km 114, waterlogged, periphyton, 26-VII-2000, *S.M.M. Faustino & S.P. Schetty*, 23°51'11,3"S, 49°09'10,8'W, conductivity 10 qS cm^{-1}, pH 6,9 (SP355392). **Municipio de Sao Paulo**, Parque Estadual das Fontes do Ipiranga, Lago do IAG, 06-III-1999, *I.S. Vercellino*, 23°38'08"S and 23°40'18"S, 46°36'48"W and 46°38'00"W (SP399780).

Comment

The genus *Pseudodidymocystis* was established by species previously described as *Didymocystis,* which have a layer of sporopollenin in the cell wall. These special structures in the cell wall resemble, but differ from, the rosettes of *Scenedesmus.* The functions of these structures are unknown, but may be the production of polysaccharides.

FOTT (1973) considered the presence or absence of pyrenoids to be a diagnostic character separating *Didymocystis* from *Scenedesmus.* Komárek (1973) highlighted the formation of two-celled cenobia as a better character and designated *D. planctonica* Korsikov as the lectotype of the genus, giving greater weight to the formation of two-celled cenobia than to the presence or absence of pyrenoids. However, COMAS (1996) mentioned that the use of the presence of pyrenoids to establish the genera of Scenedesmaceae should be combined with other characters, as there are many genera in the family whose species may or may not present pyrenoids. HEGEWALD & DEASON (1989) characterized *Scenedesmaceae* by the fact that their cells have a fine structure typical of *Scenedesmus,* but including a simple pyrenoid surrounded by starch, with three layers of sporopollenin in the cell wall.

We agree with TUCCI *ET AL.* (2006) who identified with *Pseudodidymocystis planctónica* (Korsikov) Hegewald & Deason cenobiums composed of two oblong cells arranged parallel to each other and in relation to their longitudinal axis. However, all the specimens currently studied had slightly larger individuals.

We examined representatives of P. *planctónica* (Korsikov) Hegewald & Deason from three localities in the state of São Paulo and always in small populations, consisting of only a few individuals representing the species. With regard to morphological variation, one character that was not very stable in the species was cell size, but all the other characters remained extremely stable. It was also noted that P. *planctónica* (Korsikov) Hegewald & Deason is similar in morphology to .S' *ecornis* (Ehrenberg) Chodat, however, the first species has a cell wall with characteristic granulations that do not occur in the second species.

Subfamily Scenedesmoideae *sensu* Komárek & Fott 1983

Flat, linearly arranged cenobiums, rarely semicircular cenobiums or with the cells more or less crossed; cylindrical, oval or fusiform cells joined in parallel in single or double row cenobiums, with their longitudinal axes more or less perpendicular to the plane of the cenobium, sometimes forming tetrads joined by their convex margins; parietal chloroplastidium, with or without pyrenoid; smooth cell wall. In some species, pleat-like structures form on the cell wall. Reproduction exclusively by autospores, formed inside the mother cell, which are released by breaking the wall.

Key to the identification of Scenedesmoideae genera:
1. Cenobium with 2-16(-32) cells arranged in 1-2 series, the longest axes parallel to each other, cells do not form a system of folds in the cell wall..*Scenedesmus*
1. Cenobium with (2-)4-8 cells arranged in 1 series, the axes
longer ones parallel to each other, slightly inclined towards each other.
to each other; cells forming a system of folds in the cell wall................................*Enallax*

Scenedesmus **Meyen 1829**

This genus included all green, autosporic coccoid algae that formed more or less linear cenobia; the cells could be distributed in a single plane, which could also be curved.

SMITH (1916) published the first monograph based on cultures of species from the United States of America. Something similar was done by CHODAT (1926) on material from Swiss species. These two monographs were not widely used because the culture media used was extremely concentrated and was therefore responsible for the appearance of an excessive number of anomalous forms. Worthy of mention is the work by UHERKOVICH (1967) based on material from Hungary and that by KOMÁREK & FOTT (1983) which incorporated material from all over the world. Finally, the catalog published by HEGEWALD & SILVA (1988), which includes around 1,300 taxa, i.e. all the taxa described at the time.

CHODAT (1926) established a system of four subgenera for *Scenedesmus,* two of them divided into series, which reflected the taxonomic relationships between the numerous taxa described *(Euscenedesmus* Chodat with the series *Seriad, Fasciculati, Catenati* and *Conniventes, Rhynchodesmus* Chodat, *Desmodesmus* Chodat with the *Obtusi, Denticulati, Striati* and *Quadrispinosi series* and *Clathrodesmus* Chodat, which included only two species with no definitive taxonomic status, S. *costatus* Schmidle and S. *coelastroides* Schmidle). HEGEWALD (1978) reduced the number of subgenera to three, namely: *Scenedesmus, Acutodesmus* Hegewald and *Desmodesmus* Chodat, a system endorsed by KOMÁREK & FOTT (1983).

At the end of the 1980s, a trend towards adopting a broader species concept for *Scenedesmus* could already be observed in taxonomic literature (COMAS 1996), as can be seen, for example, from the extensive list of synonyms for S. *obtusus* Meyen in HEGEWALD *ET AL.* (1988) and HINDÁK (1990). Genetic-molecular bases have led to a separation of *Senedesmus* species in a narrower sense, grouping those that are morphologically similar, as was the case with S. *ovalternus* Chodat and S. *obtusus* Meyen (HEGEWALD & HANAGATA 2000).

AN *ET al.* (1999) concluded that, on the basis of ultrastructural information and genetic-molecular data, the genus *Scenedesmus* could be divided into two independent genera: *Scenedesmus 'sensu stricto',* whose cell wall is composed of three layers of sporopollenin and has no ornamentation; and *Desmodesmus,* whose cell wall is composed of four layers of sporopollenin and has ornamentation on the outermost layer of the wall. However, there were no other criteria to reinforce this proposition and the subgenus *Acutodesmus* (includes *Tetradesmus)* was elevated to a genus by Tsarenko *in* TSARENKO & PETLEVANNY (2001).

Individuals in which the cenobiums can be flat, in series or alternating, made up of 2-16 cells (rarely 32), with the longest axes parallel to each other. The arrangement of the cells in the cenobia can be in one or two series. The cells may be ellipsoid, ovoid, fusiform or lunate and may all be the same in the same cenobium or the outer cells may be one shape and the inner cells another. The poles can be broadly rounded or attenuated, truncated or pointed. The cell wall is smooth in most species, but can also be decorated with small warts. Reproduction is carried out by 2-8 or more autospores that will form a single autocenobium still inside the mother cell, which will be released by breaking the mother cell wall. The chloroplastid is unique per cell, occupies a parietal position and has a pyrenoid.

Key to identifying the subgenera studied:
1. Fusiform or elliptical-fusiform cells, with more or less acute .... polesSubgenus *Acutodesmus*
1. Cells with rounded and bare poles

.... of thornsubgenus *Scenedesmus*

Subgenus *Acutodesmus* **Hegewald**

The subgenus *Acutodesmus* Hegewald includes *Scenedesmus* species with fusiform or elliptical-fusiform cells and more or less acuminate poles. The cell wall has no ornamentation or spines. *Scenedesmus acutiformis* Schroder, a typical species due to the presence of costeloid structures in the cell wall produced by folds in the outer layers, is currently considered to be in the genus *Enallax* (HINDÁK 1990, HEGEWALD & HANAGATA 2000).

The subgenus shows wide morphological variability in both the shape of the cells and the cenobiums (aligned, alternate, in one or two rows). This variability has led to the description of numerous taxa at both specific and subspecific levels. It has been proven that there are popular species with a wide range of morphological variation, as well as intermediate individuals within the same species, which only proves the influence of environmental factors on morphological characters, thus leading to the development of certain morphotypes (HOLTMANN & HEGEWALD 1986, MLADENOV & FURNADZIEVA 1995,1999, BELKINOVA & MLADENOV 2000, 2002). Furthermore, the generally accepted species are also subject to the criteria of different authors (HEGEWALD 1979, OOSHIMA 1981, KOMÁREK & FOTT 1983,

COMAS & KOMÁREK 1984, KRIENITZ 1987, HINDÁK 1990).

Tsarenko in TSARENKO & PETLEVANNY (2001) elevated the subgenus *Acutodesmus* to the category of genus and made the appropriate nomenclatural changes to some species previously in *Scenedesmus*. HEGEWALD & WOLF (2003) confirmed, after studies of gene sequences using 18S rDNA and ITS-2 as genetic markers, that the relationships between *Scenedesmus* and *Acutodesmus* are not entirely clear, especially in the cluster formed by A. *pectinatus* (Meyen) Tsarenko, *A. regularis* (Svirenko) Tsarenko and .S' *arcuatus* Lemmermann, which are phylogenetically related but separated from other species by the aforementioned authors (HEGEWALD & WOLF 2003). This group of species was listed as "Scenedesmaceae incertae sedis", thus not confirming the genus *Acutodesmus in* the sense of TSARENKO & PETLEVANNY (2001).

According to the above criteria, the difference between *Scenedesmus* and *Acutodesmus* is, so far, eminently morphological and not all species morphologically similar to *Acutodesmus* have been studied (S. *incrassatulus* Bohlin, S. *bourrellyi* Iltis, S. *ginzbergeri* Kammer, etc.). In these circumstances, we prefer to consider *Acutodesmus* a subgenus.

Key for identifying the species and varieties studied:
1. Cells arranged linearly or alternately2
2. Cells arranged crosswise3
3. Linearly arranged cellsS .... *regularis*
4. Cells arranged alternately, rarely linearly7
5. Cells arranged crosswiseS .... *wisconsinensis*
6. Cells not arranged crosswise4
7. Cells with similar thickening at the apex
    to a crownS .... *indicus*
8. Cells with acuminate poles, without thickening at the apex5
9. Cells arranged in series, the pole of one cell touching the subapical part of the other cell 6
10. Cells with more or less stellate arrangementS .... *acuminatus* var. *elongatus*
11. Cell 23.7-29.1 pm long, 3.6-4 pm wideS........................... *javanensis var.javanensis*
12. Cell 39.0-50.0 pm long, 7.0-8.0 pm wideS........................ *javanensis* var. *schroeteri*
13. Cell with mucron poles, obvious mucronS... . *incrassatulus*
14. Cells with acuminate or acute poles8
15. Spindle-shaped cells, acuminate polesS ... . *acuminatus* var.   *acuminatus*
16. Fusiform to oval fusiform cells, poles acuteS........................ *obliquus* var. *dimorphus*

Scenedesmus acuminatus (Lagerheim) Chodat var. *acuminatus* (Fig. 118-119)
Algues vertes de Suisse. 211. 1902.
Basionimo: *Selenastrum acuminatum* Lagerheim,   OfversigtafKungliga
Vetenskapsakademiens forhandlingar 39(2): 71,pl. 3, figs. 27-30. 1882.

Cenobium flat, formed by 4-8 cells; cells arranged linearly or alternately, fusiform, lunate, acuminate, acuminate poles, outer cells markedly arched, inner cells almost straight; cell length 20.4-34.1 pm, width 3.0-5.42 pm; parietal chloroplastid, 1 pyrenoid.
Geographical distribution in the State of Sao Paulo

IN LITERATURE: **Municipios of Americana, Atibaia, Bauru, Conchal, Guaratinguetá, Ibiúna, Irapuá, Itapeva, Jaú, Juquiá, Pontes Gestal, Registro, Santo André, Sao Bernardo do Campo, Sao Carlos, Municipio de Santo André** [SANT'ANNA 1984: 191, fig. 118, as *Scenedesmus acuminatus* (Lagerheim) Chodat var. *bernardii* (G.M. Smith) Dedusenko]. **Municipio de Juquiá** [SANT'ANNA *ET AL.* 1988: 91, fig. 39, as *Scenedesmus acuminatus* (Lagerheim) Chodat]. **Municipality of Ribeirão Preto** [SILVA 1999: 289, fig. 59, as *Scenedesmus acuminatus* (Lagerheim) Chodat]. **Municipality of Santo André** [SANT'ANNA 1984: 190, fig. 117, as *Scenedesmus acuminatus* (Lagerheim) Chodat var. *acuminatus* f. *maximus* Uherkovich]. **Municipio de Sao Carlos** [SCHWARZBOLD 1992: 112, fig. 5, as *Scenedesmus acuminatus* (Lagerheim) Chodat]. **Municipio de Sao Paulo** [HINO & TUNDISI 1977: 84, fig. 98, as *Scenedesmus acuminatus* (Lagerheim) Chodat; SANT'ANNA *et al.* 1989: 95, fig. 65-66, as *Scenedesmus acuminatus* (Lagerheim) Chodat; FERRAGUTETAL. 2005: 155, fig. 88, as *Scenedesmus acuminatus* (Lagerheim) Chodat var. *acuminatus]*. **Municipios de Sao Paulo, Sumaré e Tambaú** [SANT'ANNA 1984: 186, fig. 113-116 as *Scenedesmus acuminatus* (Lagerheim) Chodat var. *acuminatus* f. acuminatus*]*. **Municipality of Teodoro Sampaio** [BICUDO *ET AL.* 1992: 301, fig. 41, as *Scenedesmus acuminatus* (Lagerheim) Chodat var. *acuminatus* f. acuminatus*]*. **Municipality of Teodoro Sampaio** [BICUDO *ET AL.* 1992: 301, fig. 41, as *Scenedesmus acuminatus* (Lagerheim) Chodat var. *acuminatus* f. acuminatus*]*.

EXAMINED MATERIAL: **Municipality of Tambaú**, Clube de Tambaú, dam, 23- VI-1973, *D.M. Vital* (SP113574). **Municipality of Pirassununga**, SP-225, km 23, Cascalho neighborhood, lake, ?-XII-1973, *P.A.C. Senna* (SP123900).

Comment

HEGEWALD (1979) identified as S. *acuminatus* (Lagerheim) Chodat those populations in which the cells are joined to each other strictly by the convex faces, arranged in different planes, in agreement with its basionym *Selenastrum acuminatum* Lagerheim. There are other populations, however, whose cenobiums have markedly alternating cells in a single plane, which should be identified with *Scenedesmus falcatus* Chodat. Subsequently, this last morphological type was called *.S' pectinatus* Meyen 1829 by HOLTMANN & HEGEWALD (1986) based on the priority of publishing this name and the illegitimacy of S. falcatus *(S. falcatus* Chodat 1926 is a later homonym of S. *falcatus* Chodat 1895, also an illegitimate name according to HEGEWALD & SILVA (1988). The name S. *pectinatus* Meyen *sensu* Holtmann & Hegewald (1986) has been used by several authors, including MLADENOV & FURNADZIEVA (1999), after studying the morphological variability and proving the differences between the two species. According to the nomenclatural type (MEYEN 1829: figs. 33-35), S. *acuminatus* (Lagerheim) Chodat is closely related to S. *obliquus* Turpin var. *dimorphus* (Turpin) Hansgirg [= S. *dimorphus* (Turpin) Kützing] and has even been considered a synonym of the latter species (KOMÁREK & FOTT 1983). HOLTMANN (1994 cited by HEGEWALD & HANAGATA 2000) stated that in the 8-celled cenobia of S. *pectinatus* Meyen, the outer cells may be larger than the inner cells. These authors also stated that S. *pectinatus* Meyen is not morphologically very different from other *Acutodesmus* species and that it is especially difficult to distinguish it from S. *obliquus* Turpin.

However, HEGEWALD & HANAGATA (2000) differentiate these two species by the relatively longer cells which regularly form eight-celled cenobia and by the wall of their cells which may contain several rib folds similar to those of *Enallax acutiformis* (Schroder) Hindák. Overall, they indicate that genetic-molecular studies have so far not allowed us to fully understand the different phylogenetic lines within the *Acutodesmus* subgenus.

BELKINOVA & MLADENOV (2000) compared cultures with different nutrient contents and temperatures, concluding that the concentration of nutrients influenced the curvature of the marginal cells of the cenobium more in S. *pectinatus* Meyen than in S. acuminatus (Lagerheim) Chodat. *Acuminatus* (Lagerheim) Chodat. However, in the outer cells they observed a continuous gradient of variation in the two species, ranging from lunate to fusiform cells, with no overlap between the shapes, and therefore concluded that they can serve as a good diagnostic character.

The two morphotypes coexist with their different morphological characters well defined, but so far a satisfactory taxonomic solution, and especially the nomenclatural one, has not been reached. Several experts continue to identify as S. *pectinatus* Meyen *sensu* HOLTMANN & HEGEWALD (1986) an alga that should be called S. *acuminatus* (Lagerheim) Chodat *"sensu lato",* including the aforementioned morphotypes and their infra-specific categories, since many varieties and taxonomic forms currently described are based on anomalies or constitute part of the spectrum of morphological variation of a single taxon.

Slightly alternating lunate cells joined by 1/5 of the cell plane. The inner cells are fusiform and less arched than the outer cells. *Scenedesmus acuminatus* (Lagerheim) Chodat is, according to NOGUEIRA (1991), a species with numerous morphological expressions described as various taxonomic categories. We agree with NOGUEIRA (1991), as this is a very morphologically variable species which, according to COMAS (1996), includes numerous taxa at infra-specific levels. It is absolutely necessary, not only for this species, to carry out a population analysis and, in parallel, to compare it with the original diagnosis and/or description in order to further clarify the species.

Scenedesmus acuminatus (Lagerheim) Chodat has only been documented in two localities in the state of Sao Paulo. It is a relatively easy species to identify taxonomically, however, caution should be exercised regarding the similarity with morphotypes of S. *obliquus* (Turpin) Kützing var. *dimorphus* (Turpin) Kützing, as S. *acuminatus* (Lagerheim) Chodat has fusiform or lunate cells, with acuminate poles, arranged linearly or alternately, which are diagnostic characters of the species. As for morphological variation, the species only varied in terms of the shape of the cells, which were sometimes more arched (with a greater concavity) and sometimes less arched. Finally, it is interesting to note that specimens that can be identified with S. *acuminatus* (Lagerheim) Chodat types *"pectinatus"* and *"dimorphus" have* been recorded in the state of São Paulo.

Scenedesmus acuminatus (Lagerheim) Chodat var. *elongatus* G.M. Smith (Fig. 120) Transactions of the American Microscopical Society 45: 189,pl. 16,fig. 13-15. 1926.

Cenobia not always in the same plane, formed by 4 cells; cells irregularly arranged, twisted, arranged more or less like a star, lunate, poles acuminate; cell length 30.0-40.0 pm, width 3.0-5.5 pm.

Geographical distribution in the state of Sao Paulo

IN LITERATURE: **Municipality of Guaratinguetá, Itú, Jaú, Sao Bernardo do Campo** and **Sao Carlos** [SANT'ANNA 1984: 192, fig. 119, as *Scenedesmus acuminatus* (Lagerheim) Chodat var. *elongatus* G.M. Smith]. **Municipio de Sao Paulo** [SANT'ANNA *ET AL.* 1989: 96, fig. 67, as *Scenedesmus acuminatus* (Lagerheim) Chodat var. *elongatus* G.M. Smith].

MATERIAL EXAMINED: only material from the literature.

Comment

According to SANT'ANNA (1984), *Scenedesmus acuminatus* (Lagerheim) Chodat var. *elongatus* G.M. Smith differs from the typical variety of the species in its irregularly twisted cells, which form a curved plate with a more or less star-shaped arrangement of cells.

***Scenedesmus incrassatulus* Bohlin** (Fig. 121-124)

Bihang till K. Svenska vetenskapsakademiens handlingar: sér. 3, 23(7): 24, pl. 1, fig. 45-51. 1897.

Cenobium flat, formed by 4-8 cells; cells isolated or arranged linearly or alternately, more or less irregular, oval-fusiform, poles mucronate, 1 mucron very evident; outer cells markedly convex, inner cells more oval or almost straight; cell length 14.1-28.9 pm, diameter 4.4-7.8 pm; parietal chloroplastidium, 1 pyrenoid.

Geographical distribution in the State of Sao Paulo

IN LITERATURE: **Municipality of Santo Anastácio** [SANT'ANNA 1984: 224, fig. 146-149, as *Scenedesmus incrassatulus* Bohlin]. **Municipality of Sao José dos Campos** (CARDOSO 1979: 89, fig. 110, as *Scenedesmus incrassatulus* Bohlin).

EXAMINED MATERIAL: **Municipality of Ibirá**, Termas de Ibirá, lake, 25-V- 1973, *D.M. Vital* (SP113429). **Municipality of Rio Claro**, Horto Florestal, lake, 31-I-1975, *O.A. da Silva* (SP123867). **Municipality of Santo Anastácio**, Santo Anastácio-Mirante do Paranapanema highway, Santo Anastácio river, 23-VIII-1973, *D.M. Vital* (SP114511). **Municipality of Sao Paulo**, Santo Amaro, pond near Av. Washington Luiz n⁰ 3669, 01-VI-1967, *B. Skvortzov* (SP104098).

Comment

Scenedesmus incrassatulus Bohlin is a fairly well circumscribed and defined species. It is considered to be easy to recognize and identify, although it is not a very common species in the waters of the State of São Paulo. With regard to morphological variation, very different cell dimensions were obtained from the current specimens, which were either less long or less wide when compared to those in the literature. All other characters remained stable in the species.

CARDOSO (1979) mentioned that *Scenedesmus incrassatulus* Bohlin is a very characteristic species due to the morphology of the cenobia, which give the impression of being spherical. The outer cells are curved, giving the impression of being smaller, and the inner cells are straighter, giving them a globoid appearance. The original description of the species in BOHLIN (1897) mentioned that the specimens formed only 4-celled cenobia (never 2-celled), however, SANT'ANNA (1984) found individuals with cenobia formed by two and four cells. In addition, the latter author also claimed to have recorded three- and eight-celled cenobia. Polar thickening was also mentioned, a characteristic that can vary within the species, sometimes being seen and sometimes not. PHLIPOSE (1967) and SMITH (1920) mentioned the similarity between *Scenedesmus incrassatulus* Bohlin and *Scenedesmus arcuatus* Lemmermann var. *capitatus* G.M. Smith, which is actually the species S. *incrassatulus* Bohlin, of which 5. *arcuatus* Lemmermann var. *capitatus* G.M. Smith is a synonym.

***Scenedesmus indicus* Philipose *ex* Hegewald, Engelberg & Paschma** (Fig. 125)

Nova Hedwigia 47(3-4): 515, fig. 167. 1988 [= *Scenedesmusproducto-capitatus* Schmulavar. *indicus* (Philipose) Hegewald, 1976 *nomen invalidum[*

Cenobia flat, formed by 4 cells, cells arranged alternately, lunate, with thickening at the apices like a crown; outer cells markedly concave, inner cells touching sub-apically by the poles of the outer cells; cell length 12.5-13.0 pm, width 3.5-5.0 pm; chloroplastidioparietal, 1 pyrenoid.

Geographical distribution in the state of Sao Paulo

IN LITERATURE: **Municipio de Sao Paulo** (TUCCI 2002: 243, fig. 48, as *Scenedesmus indicus* Philipose; TUCCI *ET AL.* 2006: 163, fig. 49, as *Scenedesmus indicus Philipose*).

MATERIAL EXAMINED: only material from the literature.

Comment

According to HEGEWALD & SILVA (1988), S. *indicus* Philipose *ex* Hegewald, Engelberg & Paschma cells are organized in an alternating series, with the poles of the inner cells in contact with the median part of the outer cell and the end of the inner part almost entirely free. We agree with TUCCI ET AL. (2006) when they say that S. *indicus* Philipose *ex* Hegewald, Engelberg & Paschma has cenobia with four lunate cells arranged in an alternating series and thickening at the apices.

Scenedesmus indicus Philipose *ex* Hegewald, Engelberg & Paschma differs from S. *bacillaris* Gutwinski (= S. *producto-capitatus* Schmula) in that it is typical, fundamentally, for its markedly alternating cenobia, a very variable character, with intermediate individuals, which has led COMAS *ET AL.* (2007) to consider that it is a single species, but with two different varieties.

Scenedesmusjavanensis R. Chodat var. *javanensis* (Fig. 126)

Zeitschrif für Hydrologie 3: 157, fig. 47. 1926.

Synonyms: *Scenedesmus obliquus* (Turpin) Kützing f. *magnus* Bernard

  Scenedesmus bernardii G.M. Smith

  Scenedesmus acuminatus var. *bernardii* (G.M. Smith) Dedusenko

  Scenedesmus acuminatus var. *javanensis* (Chodat) Couté & Rousselin

Cenobium flat, formed of 4 cells; cells arranged alternately, lunate, asymmetrical, poles acuminate; outer cells markedly concave, inner cells asymmetrical, pole of one cell touching the subapical part of the other cell; cell length23.7-29.1 pm, width 3.6-4.0 pm; chloroplastidioparietal, 1 pyrenoid.

Geographical distribution in the State of Sao Paulo

IN LITERATURE: **Municipality of Moji-Gua^u** (MARINHO 1994: 50, fig. 9, as *Scenedesmus javanensis* R. Chodat). **Municipality of Ribeirao Preto** (SILVA 1999: 291, fig. 63, as *Scenedesmus javanensis* R. Chodat). **Municipality of Sao Paulo** [SANT'ANNA *ET AL.* 1989: 96, fig. 71, as *Scenedesmus acuminatus* (Lagerheim) Chodat var. *bernadii* (G.M. Smith) Dedusenko].

EXAMINED MATERIAL: **Municipality of Novo Horizonte**, SP-304, km 451, pond, periphyton collected with net, 14-II-2001, *C.E.M. Bicudo, S.M.M. Faustino & L.R. Godinho* (SP336349). **Municipio de Sao Paulo**, Parque Estadual das Fontes do Ipiranga, Lago das Garbas, 17-I-2007, *M. Borduqui*, 23°38'08"S e 23°40'18"S, 46°36'48"W e 46°38'00"W (SP399782).

Comment

Scenedesmus javanensis R. Chodat has cells arranged alternately in the cenobium, the cells are lunate, asymmetrical and the poles are acuminate. The outer cells are markedly concave and the inner cells asymmetrical, as well as being in contact with the sub-apical part of the other cell.

The species is typical for its spindle-shaped, slender, highly alternate cells, forming a zigzag cenobium. This arrangement of the cells of the cenobium always in the same plane has been observed in members of the groups presented above, which is why many authors do not attribute taxonomic value to this character in order to differentiate species.

When they do, it is to separate infraspecific taxa, especially from .S' *acuminatus* (Lagerheim) Chodat or S. *pectinatus* (Meyen) Tsarenko (see HOLTMANN & HEGEWALD 1986; HEGEWALD & SILVA 1988). However, BELKINOVA & MLADENOV (2000, 2002) recognized the differences between S. *bernardii* G.M. Smith and S. *pectinatus* (Meyen) Tsarenko and considered them independent species.

The taxonomy of this group is somewhat complicated. Its history and proposed nomenclatural solutions can be found in COMAS & KOMÁREK (1984). Specifically, there are two main morphotypes of *Scenedesmus javanensis* R. Chodat. The first of the two has zigzag, irregular cenobia and broadly fusiform cells, with short, acute poles. This group is known as S. *bernardii* G.M. Smith. According to COMAS & KOMÁREK (1984), the name S. *bernardii* G.M. Smith was given by SMITH (1916) to ensure its identity with S. *obliquus* (Turpin) Kützing f. *magnus* Bernard. However, it was applied to a different morphological type from BERNARD's (1908) material. Therefore, because S. *bernardii* G.M. Smith has been used in different ways, the new name of S. *pseudobernardii* Comas & Komárek was proposed for S. *bernardii* G.M. Smith. HOLTMANN & HEGEWALD (1986) proposed a somewhat unusual nomenclatural solution. According to the above authors, as S. *bernardii* G.M. Smith has no type material, they proposed its diagnosis (Smith 1916) as a holotype, including S. *obliquus* (Turpin) Kützing f. *magnus* Bernard, the basionym of the species, only as a synonym. They therefore rejected the combination S. *pseudobernardii* Comas & Komárek and proposed the new combination S. *pectinatus* (Meyen) Tsarenko var. *bernardii* Dedusenko.

The second morphotype has spindle-shaped, elongated cells with gradually attenuated, elongated and pointed cell poles. The cenobia are zigzag-shaped and regular. This morphological type corresponds exactly to S. *obliquus* (Turpin) Kützing f. *magnus* Bernard, considered an independent species and different from S. *bernardii* (G.M. Smith) Dedusenko by CHODAT (1926), to which he applied the name S. *javanensis* R.Chodat, as the epithet *magnus* had already been used previously for another *Scenedesmus*. This name has been generally accepted and always applied in this sense. HUBER-PESTALOZZI (1929) described *S. schroeteri* (Huber-Pestalozzi) Comas & Komárek, which was considered a synonym for *S. javanensis* R. Chodat by COMAS & KOMÁREK (1984) and, later, in terms of the taxonomic form of the same species (COMAS & KOMÁREK *in* TOLEDO & COMAS 1988), as it differed from the typical form of the species, mainly due to the larger size of the cells.

Scenedesmus javanensis R. Chodat is a very common species with a wide geographical distribution in Brazil: Rio de Janeiro, Sao Paulo, Goiania, Manaus (Augusto Comas, personal report, February 16, 2009).

Representatives of this species were only found in two localities in the state of Sao Paulo. It was considered an easily identifiable species due to the lunate, asymmetrical cells with acuminate poles, arranged alternately in the cenobium. As for morphological variation, only cell dimensions varied, while all other characters remained stable in the populations examined.

Scenedesmus javanensis R. Chodat var. *schroeteri* (Huber-Pestalozzi) Comas & Komárek (Fig. 127)
In: Toledo & Comas, Acta Botánica Cubana 57: 7, fig. 5. 1988.
Synonym: *Scenedesmus schroeteri* Huber-Pestalozzi

Cenobium flat, formed of 4-8 cells; cells arranged alternately, lunate, asymmetrical, poles acuminate, outer cells markedly concave, inner cells asymmetrical, the pole of one cell touching the sub-apical part of the other cell; cell length 39.0-50.0 pm, width 7.0-8.0 pm; chloroplastidioparietal, 1 pyrenoid.

Geographical distribution in the State of Sao Paulo

IN LITERATURE: **Municipio de Sao Paulo** [Sant'Anna *et al.* 1989: 96, fig. 70, as *Scenedesmus acuminatus* (Lagerheim) Chodat f. *maximus* Uherkovich].

MATERIAL EXAMINED: only material from the literature.

Comment

Scenedesmus javanensis R. Chodat var. *schroeteri* (Huber-Pestalozzi) Comas & Komárek differs from the typical variety of the species by the much larger cell sizes.

Scenedesmus obliquus (Turpin) Kützing var. *dimorphus* (Turpin) Hansgirg (Fig. 128137)
Archiv der naturwissenschaft Landesdurchf. von Bohmen 6(5): 116. 1888.
Basionimo: *Achnanthes dimorphus* Turpin, Mémoires du Museum national d'Histoire naturelle 16: 313,p. 13,fig. 12. 1828.
Synonym: *Scenedesmus dimorphus* (Turpin) Kützing
    Scenedesmus acutus Meyen

Cenobium flat, formed by 4-8 cells; cells arranged linearly or alternately, fusiform to oval-fusiform, poles acute, outer cells markedly concave and can reach straight or slightly convex, inner cells slightly convex to almost straight; cell length 6.5-31.0 pm, width 1.9-6.7 pm; parietal chloroplastidium, 1 pyrenoid.

Geographical distribution in the State of Sao Paulo

IN LITERATURE: **Municipalities of Arujá, Cananéia, Guaratinguetá, Ibirá, Itú, Jaú, Pindamonhangaba, Pirassununga, Porangaba, Ribeirão Branco, Santo André, Sao Bernardo do Campo, São Carlos, Sorocaba** and **Sumaré** (SANT'ANNA 1984: 193, fig. 120121, as *Scenedesmus acutus* Meyen var. *acutus* f. acutus*)*. **Municipalities of Itú, Ribeirão Branco, Santo André, São Bernardo do Campo, São Carlos** and **Sorocaba** (SANT'ANNA 1984: 197, fig. 122-123, as *Scenedesmus acutus* Meyen var. *acutus* f. *alternans* Hortobágyi). **Municipios of Ribeirão Branco, Santo André, São Bernardo do Campo** and **São Carlos** [SANT'ANNA 1984: 198, fig. 124-125, as *Scenedesmus acutus* Meyen var. *acutus* f *costulatus* (Chodat) Uherkovich]. **Municipio de São José dos Campos** [Cardoso 1979: 72, fig. 96-97, as *Scenedesmus acutus* Meyen var. *acutus* f. acutus*;* 76, fig. 98, as *Scenedesmus acutus* Meyen var. acutus f. *alternans* Hortobágyi; 77, fig. 99, as *Scenedesmus acutus* Meyen f. *tetradesmiformis* (Wolszynska) Uherkovich]. **Municipio de São Paulo** (SANT'ANNA ET AL. 1989: 96, fig. 72, as *Scenedesmus acutus* Meyen; SANT'ANNA ET AL. 1989: 96, fig. 73, as *Scenedesmus acutus* Meyen var. *acutus* f *alternans* Hortobágyi; SCHETTY 1998: 12, fig. 31, as *Scenedesmus acutus* Meyen var. acutus f. acutus*;* FERRAGUT ETAL. 2005: 155, fig. 89, as *Scenedesmus acutus* Meyen var. acutus f. acutus*;* 155, fig. 90, as *Scenedesmus acutus* Meyen var. *acutus* f *alternans* Hortobágyi; 155, fig. 92, as *Scenedesmus dimorphus* (Turpin) Kützing; 155, fig. 96, as *Scenedesmus obliquus* (Turpin) Kützing).

EXAMINED MATERIAL: **Municipality of Barretos**, Barretos, in the city, lake region, with grasses and Cyperaceae, 28-II-1990, *L.H.Z. Branco* (SP255772). **Municipality of Braganra Paulista**, 3 km northeast of the city of Braganca Paulista, pond, 20-VI-1973, *DM. Vital* (SP113524). **Municipality of Campos de Jordão**, Alagoinha neighborhood, lake, 11-III-1973, *M.M. Sakane* (SP130445). **Municipality of Guaratinguetá**, Clube dos 500, lake, 01-IV-1966, *C.E.M. Bicudo* (SP96965). **Municipality of Itu, SP-**

280, km 77, lake, 11-V-1977, *C.R. Leite* (SP139733). **Municipality of Jacupiranga**, SP-139, km 23, pond with cattails, 13-IX-2000, *C.E.M. Bicudo, L.A. Carneiro & S.M.M. Faustino* (SP336346). **Municipality of Juquiá**, BR-116, km 165, lake, 01-III-1973, *CEM. Bicudo & L. Sormus* (SP113664); km 160, flooded, 01-III-1973, *C.E.M. Bicudo & L. Sormus* (SP113672). **Municipality of Piedade**, SP-232, km 132.3, dammed stream, 12.8 km before the Piedade-Ibiúna junction, periphyton, 30-XII- 1991, *C.E.M. Bicudo & D.C. Bicudo* (SP255766). **Municipio de Pindamonhangaba**, BR-116, km 270-271, lagoon, 21-V-1966, *C.E.M. Bicudo* (SP96949). **Municipality of Pirassununga**, SP-225, km 23, Cascalbo neighborhood, lake, ?-XII-1973, *P.A.C. Sema* (SP123900). **Municipality of Pradópolis**, SP-291, town of Pradópolis, sweet stream, 15-VIII-2000, *C.E.M. Bicudo, S.M.M. Faustino & L.L. Morandi* (SP365701). **Municipality of Reginópolis**, SP-331, km 115.2, on the left, towards Pirajuí, on the right-hand side of the Batalba river, 500 m after the entrance to Reginópolis, marsh with macrophytes, periphyton, 22-II-1992, *C.E.M. Bicudo & D.C. Bicudo* (SP255769). **Municipio de Ribeirao Branco**, no precise locality indicated, 19-V-1972, *D.M. Vital* (SP130427). **Municipality of Rio Claro**, SP-310, km 156, flooded, 10-V-1973, *C.E.M. Bicudo & P.A.C. Senna* (SP104728). **Municipality of Rifaina**, Rio Grande bridge linking the municipalities of Rifaina and Araxá, phytoplankton, 30-V-2000, *C.E.M. Bicudo & D.C. Bicudo* (SP336342). **Municipio de Rincao**, SP-257, km 11, stream, 15-VIII-2000, C.E.M. Bicudo, *S.M.M. Faustino & L.L. Morandi* (SP365700). **Municipality of Sao Bernardo do Campo**, Rio Grande, Billings reservoir, 05-X-1972, *C.R. Leite* (SP130432). **Municipality of Sao Carlos**, SP-310, km 222, Aldeia Conde do Pinbal, creek, 10-V-1973, *C.E.M. Bicudo & L. Sormus* (SP104723). **Municipio de Sao Paulo**, Horto Florestal, lake, 16-VII-1962, *C.E.M. Bicudo & R.M.T. Bicudo* (SP96850); Santo Amaro, lake near Av. Wasbington Luiz n⁰ 3669, 01-VI-1967, *B. Skvortzov* (SP104098); Parque Estadual das Fontes do Ipiranga, Lago das Garbas, 12-VII-2006, M. *Borduqui,* 23°38'08"S and 23°40'18"S, 46°36'48'W and 46°38'00"W (SP399781); 17-I-2007, 23°38'08"S and 23°40'18"S, 46°36'48'W and 46°38'00'W (SP399782); Lago das Ninféias, 03-VIII-2007, T.R. *Santos,* 23°38'08"S and 23°40'18"S, 46°36'48"W and 46°38'00"W (SP399783). **Municipality of Sorocaba**, Brigadeiro Tobias, lake, 19-V-1972, *D.M. Vital* (SP130425). **Municipality of Tambaú**, Clube de Tambaú, dam, 23- VI-1973, *D.M. Vital* (SP113574). **Municipality of Tupa**, no precise location indicated, 20- VII-1973, *D.M.* Vital (SP130789).

Comment

Scenedesmus obliquus (Turpin) Kützing belongs to one of the most difficult to identify species groups of the *Acutodesmus* subgenus due to its wide morphological variability. The three main morphological types that have served to establish species or varieties of *Acutodesmus* are: (7) cenobia formed by broadly fusiform or fusiform-elliptical cells, more or less straight, poles acuminate-rounded, cells distributed alternately in one or two rows (type S. *obliquus),* (2) cenobia formed by cells from narrowly to broadly fusiform, outer cells clearly lobed or with a convexity on the outer margin which does not overlap the line of the poles, poles pointed, cells aligned or slightly alternating in a row (type .S' *dimorphus),* and *(3)* cenobia formed by broadly fusiform cells, slightly curved outer cells, facing outwards (at least their ends), free margins with a convexity that usually overlaps the line of the poles, more or less pointed poles, aligned or slightly alternating cells, usually in a rare two rows (type S. *acutus).* Within each of these morphotypes, different and particular modifications have been found which have given rise to numerous infraspecific taxa (Hegewald & Silva 1988).

Scenedesmus acuminatus (Lagerheim) Chodat *'sensu lato'* can be interpreted as clearly consisting of two morphotypes, *acuminatus* and *pectinatus,* without intermediate populations, which would taxonomically constitute a single species, S. *obliquus* (Turpin) Kützing, a species that would include S. *acutus* Meyen and S. *dimorphus* Turpin. This is what HEGEWALD (1979, 1989) and HOLTMANN & HEGEWALD (1987) did, while others preferred to consider three independent species (KOMÁREK & FOTT 1983). We accept the criteria of TOLEDO & COMAS (1988) and COMAS (1996) who considered two taxonomic varieties: S. *obliquus (*Turpin) Kützing var. *obliquus* and S. *obliquus* (Turpin) Kützing var. *dimorphus (Turpin) Kützing (*including S. *acutus* Meyen). *Scenedesmus obliquus* (Turpin) Kützing var. *dimorphus* is the species with the greatest morphological variability within the genus, even in the area studied.

Scenedesmus obliquus (Turp.) Kutzing var. *dimorphus* (Turp.) Kutzing is the species with the greatest morphological variability within the genus. In addition to the great morphological variability, the species is widely distributed in the state of Sao Paulo.

Scenedesmus obliquus (Turpin) Kützing var. *dimorphus* (Turpin) Kützing occurred in 22 municipalities in the state. The species is also considered to be cosmopolitan, with the greatest geographical distribution of all the species in this study. The species was not considered easy to identify due to the fact that it has several morphological types, as well as a large number of intermediate specimens in the populations examined.

Scenedesmus regularis Svirenko (Fig. 138)

Russkii arkhivprotistologii 3(1-2): 178, fig. II: 11. 1924.

Cenobium flat, formed by 4 cells; cells arranged linearly, fusiform, slender, joined in the median region for up to 1/3 of the length, poles attenuated, curved, facing the inside of the cenobium ñ the outer cells, practically straight ñ the inner cells; cell length 12.0-15.0 pm, width 2.5-4.0 pm; parietal chloroplastidium, 1 pyrenoid.

Geographical distribution in the State of Sao Paulo

IN LITERATURE: **Municipio de Sao Paulo** (Tuca 2002: 243, fig. 47, as *Scenedeamua regularia* Svirenko; Tucci *ET AL.* 2006: 165, fig. 50, as *Scenedeamua regularia* Svirenko).

MATERIAL EXAMINED: only material from the literature.

Comment

Scenedeamua regularia Svirenko is typical for its fusiform cells, joined together by 1/3 of their length and elongated, curved poles. In terms of cell shape, this species resembles *Scenedeamua obliquua* (Turpin) Kützing var. *dimorphua* (Turpin) Kützing and S. *pectinatua* (Meyen) Tsarenko. Genetic-molecular studies confirmed the kinship of S. *regularia* Svirenko with S. *pectinatua* (Meyen) Tsarenko and with S. *arcuatua* (Lemmermann) Lemmermann, and it was inserted by HEGEWALD & WOLF (2003) as Scenedesmaceae *"incertae aedia"*.

Tucci (2002) and Tucci *ET AL.* (2006) identified material from Parque Estadual das Fontes do Ipiranga with S. *regularia* Svirenko, with which we fully agree, as both materials agree with the original figure of the species in HEGEWALD & SILVA (1988: fig. 771).

Scenedeamuawiaconainenaia (G.M. Smith) Chodat (Fig. 139)

Monographie d'algues en culture pure. 52. 1913.

Basionimo: *Tetradeamua wiaconainenaia* G.M. Smith, Bulletin of the Torrey Botanical club 40: 76, pl. 1,fig. 1-2. 1913.

cenobium formed by 4 cells distributed in 2 planes; lunate cells, arranged 2 by 2 in different parallel planes, joined together by convex margins, arranged crosswise in apical view; cell length 8.0-15.2 pm, width 5.0-6.5 pm; parietal chloroplastid, 1 pyrenoid.

Geographical distribution in the state of Sao Paulo

IN LITERATURE: **Municipios of Rio Claro** and **Tambaú** (SANT'ANNA 1984: 256, fig. 169, as *Tetradeamua wiaconainenaia* G.M. Smith).

EXAMINED MATERIAL: **Municipality of Assis** SP-333, km 435, pond, with aquatic vegetation and taboanas banks, 21-VII-1991, MC. *Bittencourt-OUveira* (SP239089).

Comment

Tetradesmus wisconsinensis G.M. Smith is now *Scenedesmus wisconsinensis* (G.M. Smith) Chodat on the basis of molecular biology studies by HEGEWALD & HAGANATA (2000), which proved that *Tetradesmus* is actually *Scenedesmus*.

Tetradesmus is defined by the bundled arrangement of the cells in the cenobium, i.e. they are arranged two by two in two planes. The genus was described by SMITH (1913) from cultures prepared to observe morphological variation in the genus and he found a lot of similarity between some of these forms and those of *Scenedesmus acutus* Meyen, a species which, according to CHODAT (1909), shows great variability under different cultivation conditions. The regular arrangement of the cells in a linear series disappeared and the cells became either isolated or arranged in branched filamentous structures that NÁGELI (1848) described under the generic name *Dactylococcus*. CHODAT's (1909) investigations did not, however, find any variations of S. *acutus* Meyen that could be identified as *Tetradesmus*. *Tetradesmus* was considered a synonym for *Scenedesmus* by CHODAT (1926, but was revived by KOMÁREK & FOTT (1983), who added other species to it. HEGEWALD & HAGANATA (2000) included *Tetradesmus* in the synonymy of the *Acutodesmus* subgenus of *Scenedesmus*. *Scenedesmus wisconsinensis* (G.M. Smith) Chodat was only found in one municipality in the state of Sao Paulo and the population studied was quite small. It should be noted that there was no difficulty in identifying the representatives of this species by taxonomy because their diagnostic characteristics were very evident: lunate cells, arranged two by two in different planes, joined together by convex margins and arranged crosswise in apical view. As far as morphological variation is concerned, the few individuals examined kept their morphological characteristics fairly stable, especially the diagnostic ones, with the exception of cell dimensions, which fluctuated very little within the species.

Subgenus *Scenedesmus* Meyen

The subgenus comprises species with cells with rounded poles and no spines. In another layer of the cell wall formed by outer layers of sporopollenin, granules or warts produced by iron and manganese impregnations may appear. Cenobia are variable in shape, with cells appearing perfectly aligned, slightly or markedly alternating to disciform, in a spatially ordered plan. Their cell wall is made up of three layers, one of which is sporopollenin. These characteristics are traditionally used to differentiate species or infra-specific categories of this subgenus.

The taxonomy of this subgenus is especially complex, as most species show a wide variation in the characteristics considered diagnostic, which has led to the description of numerous taxa including the repetition of epithets such as *disciformis, flexuosus, platydiscus,* etc. further complicating the taxonomy of the subgenus. KOMÁREK & FOTT (1983) put the then complicated nomenclature of the subgenus in order, but HEGEWALD *ET AL.* (1988) opposed many of the solutions in KOMÁREK & FOTT (1983).

The criteria in KOMÁREK & FOTT (1983) are often based on new interpretations of the meanings of certain species or the designation of the nomenclatural types of some of them. The species were generally very broadly conceived, as is the case, for example, with S. *obtusus* Meyen and S. *arcuatus* (Lemmermann) Lemmermann. The criteria in KOMÁREK & FOTT (1983) are also reflected in HNDÁK (1990).

Molecular genetics has shown an excellent basis for separating the genera *Scenedesmus* and *Desmodesmus* (AN *ET AL.* 1999), but it has also shown surprising phylogenetic relationships, such as the proximity of S. *arcuatus* (Lemmermann) Lemmermann to S. *regularis* Svirenko and S. *pectinatus* (Meyen) Tsarenko, a group of species considered Scenedesmaceae *"incertae sedis"* (HEGEWALD & WOLF 2003).

Key to identifying species and varieties:
1. Cenobium formed by 2 cells, without pyrenoidS .... *bicellularis*
2. Cenobium formed by 4 or more cells, with pyrenoid2
   3. Cell wall with small wartsS .... *verrucosus*
   4. Cell wall without warts3
5. Curved, semicircular cenobiumS... *curvatus*
6. Flat cenobiums, arranged linearly or alternately4
   7. Elliptical or .... oblong cells5
   8. .... Oval-cylindrical cells6
   9. Oblong cellsS .... *ecomis*
10. ....Elliptical cells9
   11. Cells with intercellular spathesS ..... *arcuatus* var. *platydiscus*
   12. Cells without intercellular spathesS .... *obtusus*
13. Cells with thickening at the cell polesS .... *ellipticus*
7. Cells without thickening at the cell polesS.. *acunae*

***Scenedesmus acunae* Comas** (Fig. 141-143) Acta Botánica Cubana 2: 7-8, fig. 7d-f. 1980.

Cenobium flat, formed by 4-8 cells; cells arranged linearly (rarely alternating), elliptical, poles rounded, outer cells markedly convex, inner cells straight; spines absent in all cells of the cenobium; cell length 5.018.4 pm, width 2.0-6.1 pm; parietal chloroplastid, 1 pyrenoid.

Geographical distribution in the state of Sao Paulo

IN LITERATURE: **Municipality of Juquiá** [SANT'ANNA *ET AL.* 1988: 91, fig. 40, as *Scenedesmus bijugus* (Turpin) Kützing]. **Municipality of Sao Paulo**, Guarapiranga reservoir [KLEEREKOPER 1937: fig. 7, as *Scenedesmus bijuga* (Turpin) Lagerheim; XAVIERETAL. 1985: 183, fig. 75, as *Scenedesmus bijugus* (Turpin) Kützing].

EXAMINED MATERIAL: **Municipality of Bragan^a Paulista**, 3 km northeast of the city of Braganca Paulista, pond, 20-VI-1973, *D.M. Vital* (SP113524). **Municipality of Campos de Jordao**, Alagoinha neighborhood, lake, 11-III-1973, *M.M. Sakane* (SP130445). **Municipality of Cananéia**, Ilha Comprida, 120 m from the sea, lagoon, 01-III-1975, *M. Vital* (SP130813). **Municipality of Guaratingueta**, Clube dos 500, lake, 01-IV-1966, *C.E.M. Bicudo* (SP96965). **Boundary between the municipalities of Jaú** and **Bariri,** SP-304, km 317.5, 13 km before Bariri, Santa Fé farm, pond with aquatic plants, periphyton, 22-II-1992, *C.E.M. Bicudo & D.C. Bicudo* (SP255768). **Municipio de Lins,** SP-300, km 436.5, marsh, 14-VIII-2001, *C.E.M. Bicudo, L.R. Godinho & C.I. Santos,* 21°43'53.2" S, 49°42'31.9" W, pH 6.3 (SP355377). **Municipality of Moji das Cruzes,** SP-88, km 74-75, lagoon, 18-VI-1973, *C.E.M. Bicudo, C.R. Leite & L. Sormus* (SP113662). **Municipality of Pindamonhangaba,** Sao Joao farm, Sao Joao lagoon, 21-V-1966, *C.E.M. Bicudo* (SP96950). **Municipality of Pirassununga,** SP-225, km 23, Cascalho neighborhood, lake, ?-XII-1973, *P.A.C. Senna* (SP123900). **Municipality of**

Rancharia, Vila Agisse, Ribeirao Capivari, 25-VIII-1973, *D.M. Vital* (SP114513). **Municipality of Sao Carlos**, SP-310, km 222, aldeia Conde do Pinhal, stream, 10-V-1973, *C.E.M. Bicudo & L. Sormus* (SP104723). **Municipio de Sao Paulo**, Santo Amaro, lagoon near Av. Washington Luiz n⁰ 3669, 01-VI-1967, *B. Skvortzov* (SP104098); Parque Estadual das Fontes do Ipiranga, Lago do IAG, *I.S. Vercellino*, 03-VIII-1998, 23°38'08"S and 23°40'18"S, 46°36'48"W and 46°38'00"W (SP399779); I. S. Vercellino, 06-VI-1967, *B. Skvortzov* (SP399779).S. *Vercellino*, 06-III-1999, 23°38'08"S and 23°40'18"S and 46°36'48"W, 46°38'00"W (SP399780); Lago das Garcas, col. M. *Borduqui*, 12-VII-2006, 23°38'08"S and 23°40'18"S, 46°36'48"W and 46°38'00"W (SP399781); Lago das
Nymphaea, *T.R. Santos*, 03-VIII-2007, 23°38'08"S and 23°40'18"S, 46°36'48"W and 46°38'00"W (SP399783). **Municipality of Sorocaba**, Brigadeiro Tobias, lake, 19-V-1972, *D.M. Vital* (SP130425). **Municipality of Tambaú**, Clube de Tambaú, dam, 23-VI-1973, *D.M. Vital* (SP113574).
Comment

Scenedesmus acunae Comas is easily confused with *Scenedesmus ecomis* (Ralfs) Chodat and *Scenedesmus ellipticus* Corda, all three species being very similar to each other as they have very similar descriptions and morphotypes. However, S. *acunae* Comas has more arched, convex outer cells, S. *ecomis* (Ralfs) Chodat relatively more oblong cells and S. *ellipticus* Corda also has elliptical cells, but with a thickening of the cell wall. The three species are very close morphologically and difficult to separate. Specimens of this species were collected from 14 locations in the state of São Paulo. *Scenedesmus acunae* Comas is therefore a species that is well distributed geographically in the state. Numerous populations of representatives of this species were analyzed which, in terms of morphological characteristics, showed variability with regard to cell dimensions and the convexity of the free face of the outer cells of the cenobium, which sometimes appeared more or less arched.

***Scenedesmus arcuatus* (Lemmermann) Lemmermann var. *platydiscus* G.M. Smith** (Fig. 144-145)
Transactions of the Wisconsin Academy of Sciences, Arts & Letters 18: 451, pl. 30, fig. 101105, 122. 1916.

Cenobia curved, with intercellular spaces or very small spaces, formed by 4-8 cells; cells in 1-2 rows, arranged alternately, oval-cylindrical, poles rounded; outer cells slightly concave, sometimes projecting downwards, sometimes upwards, outer cells not completely aligned; cell length 5.7-8.2 pm, width 1.7-4.2 pm; parietal chloroplast, 1 pyrenoid.

Geographical distribution in the state of Sao Paulo

IN LITERATURE: **Municipalities of Itu, Jaú, Moji-Guaçu, Rio Claro, Salesópolis, Santo André** and **Ubatuba** [SANT'ANNA 1984: 212, fig. 133-135, as *Scenedesmus bijugus* (Turpin) Kützing var. *disciformis* (Chodat) Leite]. **Municipios de Itú e Santo André** (SANT'ANNA 1984: 200, fig. 126, as *Scenedesmus arcuatus* Lemmermann var. *arcuatus* f. arcuatus). **Municipality of Luiz Antonio** (PERES & SENNA 2000: 474, fig. 28, as *Scenedesmus bijugus* Chodat). **Municipality of Ribeirao Preto** [SILVA 1999: 289, fig. 60, as *Scenedesmus arcuatus* (Lemmermann) Lemmermann var. *platydiscus* G.M. Smith]. **Municipios de Sao Paulo e Itu** (SANT'ANNA 1984: 202, fig. 127-128, as *Scenedesmus arcuatus* Lemmermann var. *arcuatus Lemmermann* f. *gracilis* Hortobágyi). **Municipio de Sao Paulo** [SANT'ANNA *ET AL.* 1989: 96, fig. 74-75, as *Scenedesmus arcuatus* Lemmermann); 96, fig. 78, as *Scenedesmus bijugus* (Turpin) Kützing var. *disciformis* (Chodat) Leite].

EXAMINED MATERIAL: **Municipality of Itu**, SP-280, km 77, lake, 11-V-1977, *C.R. Leite* (SP139733). **Municipality of Pindamonhangaba**, BR-116, km 270-271, lake, 21-V-1966, *C.E.M. Bicudo* (SP96949).
Comment

This variety differs from the typical species in that its cells are all in the same plane. *Scenedesmus arcuatus* (Lemmermann) Lemmermann var. *platydiscus* G.M. Smith can easily be confused with *Scenedesmus obtusus* Meyen, however, the former has larger intercellular spathes than *Scenedesmus obtusus* Meyen, and the cells of the latter species are more uniformly fusiform, while those of S. *arcuatus* (Lemmermann) Lemmermann var. *platydiscus* G.M. Smith have slightly concave margins.

Scenedesmus arcuatus (Lemmermann) Lemmermann var. *platydiscus* G.M. Smith is characterized by its curved cenobia, made up of eight cells whose outer poles touch each other, which in the terminology of the genus is called costulate arrangement. In this respect, it differs from S. *curvatus* Bohlin, which also forms curved cenobiums, but with one of the outer cell poles free. HEGEWALD *ET al.* (1988) recognized two taxonomic varieties in S. *arcuatus* (Lemmermann) Lemmermann, namely: the typical var. *arcuatus*, which has the characteristics described above, and var. *platydiscus* G.M. Smith [= S. *platydiscus* (G.M. Smith) Chodat], which differs from the standard variety in that its cenobia are flat. While var. *arcuatus* is very variable, var. *platydiscus* G.M. Smith is much less so. According to its holotype (SMITH 1916: figs.

101-105), there is great similarity between S. *arcuatus* (Lemmermann) Lemmermann var. *platydiscus* G.M. Smith and S. *disciformis* (Chodat) Fott & Komárek (no nomenclatural type assigned). If S. *disciformis* (Chodat) Fott & Komárek is considered a good species from a taxonomic point of view, it will be necessary to designate a nomenclatural type for the aforementioned binomial to be validly published. The flat cenobium morphotype, formed by eight cells joined more or less closely together, with spaces between the rows or with tiny spaces, exists in nature and has a wide universal geographical distribution and is therefore considered cosmopolitan.

FOTT & KOMÁREK (1960) proposed .S' *disciformis* (Chodat) Fott & Komárek from S. *ecornis* (Ralis) Chodat var. *disciformis* Chodat. HEGEWALD *ET AL.* (1988) and HINDÁK (1990) showed that the basionym of S. *disciformis* (Chodat) Fott & Komárek is actually S. *verrucosus* Roll.

Considering that S. *disciformis sensu* Fott & Komárek is independent of S. *verrucosus* Roll, but morphologically very close to S. *arcuatus* (Lemmermann) Lemmermann var. *platydiscus (Lemmermann)* Lemmermann, COMAS (1996) considered the possibility that they are probably the same species. FOTT & KOMÁREK (1960) considered S. *disciformis sensu* Fott & Komárek to be a variety of S. alternans' *S. alternans* Reinsch var. *platydiscus (*G.M. Smith) Fott & Komárek, name not valid). It is not possible to state categorically that S. *arcuatus* (Lemmermann) Lemmermann var. *platydiscus* G.M. Smith and S. *disciformis sensu* Fott & Komárek are identical, since in the illustrations (holotype) in SMITH (1916) obvious spathes can be seen between the rows of cells. Therefore, we currently prefer to use the name S. *arcuatus* (Lemmermann) Lemmermann var. *platydiscus* G.M. Smith. Specimens of S. *arcuatus* (Lemmermann) Lemmermann var. *platydiscus* G.M. Smith have only been documented in two localities in the state of Sao Paulo and always in the form of populations consisting of a few individuals. The species is not easy to identify due to the similarity of its representatives with those of several species of the genus. Finally, no significant morphological variation was documented in the populations examined.

Scenedesmus bicellularis Chodat (Fig. 146-147) Preslia45' 313. 1973.
Synonym' *Didymocystis bicellularis* (Chodat) Komárek

Cenobia flat, formed by 2 cells; cells arranged linearly, elliptical-cylindrical, outer margin markedly convex, may form 4-cell cenobia, but are actually 2-cell cenobia; cell length 6.8-7.8 pm, width 2.1-3.7 pm; parietal chloroplastidium, no pyrenoid.

Geographical distribution in the state of Sao Paulo

IN LITERATURE' **Municipio de Sao Paulo** [FERRAGUT *etAL.* 2005' 154, fig. 86, as *Didymocystis bicellularis* (Chodat) Komárek].

EXAMINED MATERIAL: **Municipality of Cananéia**, Ilha Comprida, 120 m from the sea, lagoon, 07-III-1975, *D.M. Vital* (SP130813).

Comment

The genus *Didymocystis* was described by KORSIKOV (1953) and classified in the Scenedesmaceae family, including several species, but without a nomenclatural type designation. In 1973, two publications appeared, one by FOTT (1973) and the other by KOMÁREK (1973). FOTT (1973) accepted the genus with the characteristic absence of pyrenoids that differentiated it from *Scenedesmus*. The same author designated his nomenclatural type D. *inermis* Korsikov, which included D. *tuberculata* Korsikov as a synonym. *Didymocystis* species with pyrenoids were classified as *Scenedesmus*.

HEGEWALD (1978) did not accept *Didymocystis* and considered it to be synonymous with *Scenedesmus*. However, HEGEWALD (1988) accepted *Didymocytis* in the sense of FOTT (1973), studied cultures of D. *inermis* Korsikov under the electron microscope and discovered that the cell wall of this species is made up of cellulose fibers similar to those of representatives of the Oocystaceae family. Consequently, he transferred D. *inermis* Korsikov to this family. The other species of *Didymocystis* were transferred back to the genus *Scenedesmus,* including D. *bicellularis* (Chodat) Komárek, i.e. according to its original conception as *Scenedesmus bicellularis* Chodat.

HEGEWALD & DEASON (1989) described *Pseudodidymocystis* to accommodate the species previously considered to be *Didymocystis,* whose cell wall did not have cellulose fibers, but sporopollenin layers. These authors found important ultrastructural differences in the cell wall of *Scenedesmus,* especially the absence of cellulose and the presence of rosette-like structures. These substantially different structures have been given the name "bowl-shaped structures" and their presence would justify the classification in *Scenedesmus of* the species that possess them. Only *P. planctonica* (Korsikov) Hegewald & Deason (= *Didymocystis planctonica* Korsikov) and *P. fina (*Komárek) Hegewald & Deason (= *Didymocystis fina* Komárek) were considered by HEGEWALD & DEASON (1989) to be representatives of *Pseudodidymocystis,* the first of the two being the type species of the genus.

However, HEGEWALD & DEASON (1989) observed that the presence of sporopollenin caps in the studied cultures of *P. planctonica* Korsikov and *P. fina* Komárek could relate these species to representatives of the *Desmodesmus* subgenus of *Scenedesmus*. HEGEWALD & SILVA (1988) included the two species in the synonymy of *Scenedesmus bicellularis* (Chodat) Komárek and not as independent species of *Pseudodidymocystis*. HINDÁK (1990), who had previously considered *D. fina* Komárek to be *Choricystis, has* now transferred several dubious species of *Didymocystis* to *Pseudodidymocystis, such as* *D. inconspicua* Komárek, but has not taken any action with regard to *D. bicellularis* (Chodat) Komárek.

Representatives of this species were recorded in only one municipality in the state, the municipality of Cananéia. *Scenedesmus bicellularis* Chodat was not considered common in the area of the state, nor did it form populations with many individuals. Finally, it showed no significant morphological variation.

Scenedesmus curvatus **Bohlin** (Fig. 148)
Bihangtill K. Svenska vetenskapsakademiens handlingar: sér. 3, 3(7): 23, fig. 41-44, 52. 1897.

Cenobia curved, semicircular in apical view, made up of 4-8 cells; cells arranged in 2 alternating series, oblong, poles rounded; cells touching at only one of the poles; cell length 15.0-16.0 pm, width 5.0-6.9 pm. parietal chloroplastidium, 1 pyrenoid.

Geographical distribution in the state of Sao Paulo

IN LITERATURE: **Municipality of Avaré** (SANT'ANNA 1984: 214, fig. 136-138, as *Scenedesmus curvatus* Bohlin).

MATERIAL EXAMINED: only material from the literature.

Comment

According to CHODAT (1926), BOHLIN (1897) only observed populations with cenobia formed by eight cells. SANT'ANNA (1984), however, found populations with cenobia formed by four and eight cells in materials from the state of Sao Paulo.

This species was transferred to *Schroederiella* by FOTT & KOMÁREK (1960), then to *Rayssiella* by KOMÁREK (1974) and finally back to *Scenedesmus,* taking into account that *R. hemisphaerica* Edelstein & Prescott, the type species of the genus *Rayssiella,* is a synonym of *Scenedesmus arcuatus* (HEGEWALD *ETAL.* 1988).

Scenedesmus ecornis **(Ehrenberg) Chodat** (Fig. 149)
Zeitschrif fur Hydrologie 3: 170. 1926.

Basionimo: *Scenedesmus quadricaudatus* (Turpin) Ehrenberg var. *ecornis* Ehrenberg *ex* Ralfs,
Annals & Magazine ofNatural History 15: 402, pl. 12: fig. 4c. 1845.

Synonym: *Scenedesmus bijugus* (Turpin) Kützing

Cenobium flat, formed by 2-4 cells; cells arranged linearly, oblong, poles rounded, outer cells arched, markedly convex, inner cells slightly less convex; cell length 3.0-15.5 pm, width 2.0-8.0 pm; parietal chloroplastidium, 1 pyrenoid.

Geographical distribution in the State of Sao Paulo

IN LITERATURE: **Municipalities of Americana, Atibaia, Avaí, Bauru, Bragan^a Paulista, Campos do Jordao, Cananéia, Conchal, Dois Córregos, Guaratinguetá, Ibirá, Ibiúna, Itatinga, Itirapina, Itu, Jaú, Juquiá, Miracatu, Mogi das Cruzes, Moji-Gua^u, Pindamonhangaba, Pirassununga, Piratininga, Porangaba, Pontes Gestal, Rancharia, Registro, Rio Claro, Salesópolis, Santo André,** Sao **Bernardo do** Campo, **Sao Carlos,** Sao **Paulo, Sorocaba, Sumaré, Tambaú** and **Ubatuba** [SANT'ANNA 1984: 206, fig. 131-132, as *Scenedesmus bijugus* (Turpin) Kützing orth. mut. var. *bijugus].* **Municipio de Sao Paulo** [FERRAGUTET *AL.* 2005: 155, fig. 93, as *Scenedesmus ecornis* (Ehrenberg) Chodat].

MATERIAL EXAMINED: only material from the literature.

Comment

According to COMAS (1996), S. *ecornis* (Ehrenberg) Chodat has a completely smooth cell wall, with no ornamentation or apical thickening. The specimens identified in FERRAGUT *etAL.* (2005) show these same characteristics, although the specimen illustrated has only two cells in the cenobium.

Scenedesmus bijugus (Turpin) Kützing is a heterotypic synonym of *Scenedesmus ecornis* (Ehrenberg) Chodat. It is a very common species and very well represented in the state of Sao Paulo. *Scenedesmus bijugus* (Turpin) Kützing has oblong cells and cenobia consisting of two to eight cells. The specimens in SANT'ANNA *ET AL.* (1989) have only two cells in the cenobium. In this study, we consider S. *bijugus* (Turpin) Kützing to be a synonym of S. *ecornis* (Ehrenberg) Chodat, thus including the material from the state of Sao Paulo.

Scenedesmus bijugus (Turpin) Kützing had several nomenclatural problems. First, KÜTZING (1834) proposed the combination *Scenedesmus bijugatus* (Turpin) Kützing, however, some authors began to adopt *Scenedesmus bijuga* (Turpin) Lagerheim. SANT'ANNA (1984) believes that, from a morphological point of view, S. *ecornis* (Ehrenberg) Chodat, S. *bijuga* (Turpin) Lagerheim and S. *bijugatus* (Turpin) Kützing are identical species and should therefore be considered synonyms from a nomenclatural point of view. SANT'ANNA & BICUDO (1981) revised the nomenclature of S. *bijugatus* (Turpin) Kützing, resurrecting the original spelling S. *bijugus* (Turpin) Kützing and considering .S' *quadricaudatus (Turpin)* Ehrenberg and S. *ecornis* (Ehrenberg) Chodat, among others, synonyms of S. *bijugus* (Turpin) Kützing.

The name S. *bijugus* (Turpin) Kützing appears frequently in the literature, but many of the individuals identified with it actually belong to other species and generally to S. *ellipticus* Corda. *Scenedesmus minor* Kützing and S. *leibleinii* Kützing, if considered independent species, differ in cell shape and probably belong to *Desmodesmus*, as they have spines, denticles or ribs (*HEGEWALDETAL.* 1988).

KOMÁREK & FOTT (1983) accepted S. *bijugus* (Turpin) Kützing, however, HEGEWALD ETAL. (1988) did not include it in their monograph. The type species of the subgenus, S. *quadricaudatus* (Turpin) Ehrenberg var. *ecornis* Ehrenberg *ex* Ralfs, shows a cenobium whose cells resemble S. *quadrijugus* (Turpin) Trévisan, related to the subgenus *Acutodesmus*.

Although S. *bijugus* (Turpin) Kützing has been extensively recorded in the universal literature according to KOMÁREK & FOTT (1983), we believe that the existence and taxonomy of this species are still unclear, requiring further studies and more information on the species to resolve the aforementioned issues. According to RALFS (1845), there are at least four combinations to be evaluated, of which the oldest is S. *bijuga* (Turpin) Lagerheim or, more correctly, S. *bijugus* (Turpin) Kützing according to SANT'ANNA & BICUDO (1981). There are still studies to be carried out in order to define the correct binomial combination of this species, including, in this study, S. *obliquus* (Turpin) Kützing.

Scenedesmus ellipticus Rope (Fig. 150)
Almanach de Carlsbad 5: 208, pl. 4, fig. 48-49. 1835.
Synonym: *Scenedesmus linearis* Komárek 1974

Cenobium flat, formed by 4 cells; cells linearly arranged, elliptical, poles rounded, outer and inner cells thickened at the poles; cell length 7.5-8.0 pm, width ca. 2.5 pm.

Geographical distribution in the state of Sao Paulo
IN LITERATURE: **Municipio de Sao Paulo** (*FerragutetAL.* 2005: 155, fig. 95, as *Scenedesmus linearis* Komárek).

MATERIAL EXAMINED: only material from the literature.

Comment
The representatives of S. *ellipticus* Corda can easily be confused with those of *Scenedesmus ecornis* (Ehrenberg) Chodat, however, they differ due to the elliptical shape of the cells and the thickening of the cell poles in S. *ellipticus* Corda.

This species is characterized by forming cenobia with four, eight, 16 or more elliptical cells joined together in cenobia in a row, with the outer cells being slightly smaller than the inner ones. *Scenedesmus ecornis* Komárek is also morphologically very similar to this species, but is distinct because its cells do not have thickened cell walls at the poles. These thickenings are frequent in S. *ellipticus* Corda. According to HEGEWALD *ET al.* (1988) these thickenings are pure optical illusions and not real cell wall structures.

The name S. *ecornis* (Ehrenberg) Chodat has been used frequently in the taxonomic identification of specimens of this species, including numerous combinations.

Scenedesmus obtusus Meyen (Figs 151-154)
Verhandlungen der K. Leopoldinisch-carolinischen deutschen Akademie der Naturforscher 14: 775, pl. 43, fig. 30-31. 1829.

Cenobium flat, formed by 4-8 cells; cells arranged alternately in 1-2 series, oval-cylindrical, poles rounded, outer cells convex, may show 1 slight median concavity, inner cells straighter, some cells of the cenobium may show thickening of the wall; cell length 5.0-20.3 pm, width 2.0-7.45 pm; parietal chloroplastidium, 1 pyrenoid.

Geographical distribution in the State of Sao Paulo
IN LITERATURE: **Municipality of Atibaia, Cananéia, Ribeirao Branco, Salesópolis, Sao Bernardo do Campo**, Sao **Carlos**, Sao **Paulo** and **Ubatuba** (SANT'ANNA 1984: 230, fig. 153-154, as *Scenedesmus ovalternus* Chodat). **Municipality of Sao José dos Campos** [CARDOSO 1979: 86, fig. 108-109, as *Scenedesmus ecornis* (Ehrenberg) Chodat var. *ecornis].* **Municipio de Sao Paulo** [SANT'ANNA *ET AL.* 1989: 97, fig. 88, as *Scenedesmus ovalternus* Chodat; GENTIL 2000: 55, fig. 13, as *Scenedesmus arcuatus* Lemmermann; FERRAGUT *ET AL.* 2005: 156, fig. 97, as *Scenedesmus obtusus* Meyen var. *obtusus;* 156, fig. 98, as *Scenedesmus ovalternus* Chodat].

EXAMINED MATERIAL: **Municipio de Barretos**, Barretos, in the city, lake region, with grasses and Cyperaceae, 28-II-1990, *L.H.Z. Branco* (SP255772). **Municipality of Guaratinguetá**, Clube dos 500, lake, 01-IV-1966, *C.E.M. Bicudo* (SP96965). **Municipality of Piquete**, road linking Lorena to Piquete, km 65, stream, periphyton, 19-IX-2001, *C.E.M. Bicudo, D.L. Costa & F.C. Pereira*, 22°37'24.2"S, 45°09'40.1"W, pH 6.2 (SP355360). **Municipality of Rio Claro**, Horto Florestal, lake, 31-I-1975, *O.A. da Silva* (SP123867). **Município de Sao Carlos**, SP-310, km 222, aldeia Conde do Pinhal, stream, 10-V-1973, *C.E.M. Bicudo &L. Sormus* (SP104723). **Município de Sao Luiz do Paraitinga**, SP-125, km 76, pond, periphyton, 19-IX-2001, *C.E.M. Bicudo, D.L. Costa & F.C. Pereira*, GPS 23°21'58,8"S, 45°08'30,8"W, pH 6,0 (SP355363). **Município de Sao Paulo**, Santo Amaro, lagoon near Av. Washington Luiz n⁰ 3669, 01-VI-1967, *B. Skvortzov* (SP104098); Parque Estadual das Fontes do Ipiranga, Lago das Garbas, 12-VII-2006, *M. Borduqui*, 23°38'08"S and 23°40'18"S, 46°36'48"W and 46°38'00"W (SP399781); 17-I-2007,M. *Borduqui*, 23°38'08"S and 23°40'18"S, 46°36'48"W and 46°38'00"W (SP399782); Lago das Ninféias, 03-VIII-2007, *T.R.. Santos*, 23°38'08"S and 23°40'18"S, 46°36'48"W and 46°38'00"W (SP399783). **Municipality of Tambaú**, Clube de Tambaú, dam, 23-VI-1973, *D.M. Vital* (SP113574).

Comment

Scenedesmus obtusus Meyen can easily be confused with *Scenedesmus ovalternus* Chodat var. *ovalternus* and with *Scenedesmus arcuatus* Lemmermann var. *platydiscus* G.M. Smith. They differ, however, because S. *obtusus* Meyen has smaller intercellular spathes than those of S. *arcuatus* (Lemmermann) Lemmermann var. *platydiscus* G.M. Smith and its cells are relatively more fusiform, while those of S. *arcuatus* (Lemmermann) Lemmermann var. *platydiscus* G.M. Smith have a slightly concave margin.

Scenedesmus obtusus Meyen should be identified with great caution, as it is a species that can be confused with several others and is therefore very easily misidentified.

Scenedesmus ovalternus Chodat var. *ovalternus* is not considered a taxonomically valid name and should therefore not be used.

CARDOSO (1979) identified certain materials as representatives of S. *ecornis* (Ehrenberg) Chodat var. *ecornis*, however, on re-examining the illustration in that work, we concluded that it was S. *obtusus* Meyen.

Scenedesmus obtusus Meyen is the type species of the genus. The lectotypification was done by HEGEWALD *ET al.* (1975), as the original publication contained illustrations of two different species with the same name. This species currently comprises numerous taxa of infraspecific levels (HEGEWALD *ET AL.* 1988, HEGEWALD & SILVA 1988, HINDÁK 1990) and is characterized by the shape of the poles and the type of cell margin. Among its best-known synonyms are .S' *alternans* Reinsch, S. *graevenitzii* (Bernard) Chodat and S. *ovalternus* Chodat. This last name is applied, according to HEGEWALD & SILVA (1988), to a collective species and is therefore considered illegitimate according to the International Code of Botanical Nomenclature. Based on molecular biology studies, however, it is possible to separate S. *obtusus* Meyen from S. *ovalternus* Chodat. The nomenclatural solution not yet taken will be to apply a correct name to the latter species.

Representatives of S. *obtusus* Meyen were found in eight localities in the state of Sao Paulo. With regard to morphological variation, mention should be made of the wide variability in the cell dimensions of the specimens examined and the presence or absence of a thickening in the cell wall. In addition to these characters, the presence or absence of ornamentation on the cell wall should also be noted. *Scenedesmus obtusus* Meyen should be identified with caution, as its representatives can easily be confused with those of many other species, as mentioned elsewhere.

Scenedesmus verrucosus **Roll** (Fig. 155)
Russkii arkhivprotistologii4: 150. 1925.

Cenobium flat, formed by 4-8 cells; cells arranged in 2 series, approximately oblong; cell wall with small warts; cell length 6.5-10.0 pm, width 3.0-9.0 pm; parietal chloroplastidium, 1 pyrenoid.

Geographical distribution in the state of Sao Paulo

IN LITERATURE: **Municipios de Jaú e Pirassununga** (SANT'ANNA 1984: 253, fig. 167-168, as *Scenedesmus verrucosus* Roll). **Municipality of Sao Paulo** (SANT'ANNA *ET AL.* 1989: 97, fig. 96-97, as *Scenedesmus verrucosus* Roll).

EXAMINED MATERIAL: **Municipality of Cerqueira César**, SP-270, km 13, stream with strong current, plankton, 21-IX-2000, *L.L. Morandi & S.P. Schetty* (SP336348). **Município de Sao Paulo**, Parque Estadual das Fontes do Ipiranga, Lago das Garcas, 12-VII- 2006, M. *Borduqui*, 23° 38'08"S e 23° 40'18"S, 46°36'48"W e 46°38'00"W (SP399781).

Comment

Scenedesmus verrucosus Roll has cenobia made up of four or eight cells which are distributed in two rows. According to SANT'ANNA (1984), these cells are compactly arranged and appear angular due to mutual compression. They also have a cell wall decorated with small warts. UHERKOVICH (1966) considered S. *granulatus* West & West f' *disciformis* Chodat to be a heterotypic synonym of .S' *verrucosus* Roll; however, HEGEWALD & SCHNEPF (1974) concluded that the arrangement of the cells in the cenobium and the presence of warts on the cell wall are sufficient characteristics to maintain S. *verrucosus* Roll as a distinct species from S. *granulatus* West & West. However, since S. *granulatus* West & West and S. *verrucosus* Roll are easily confused from the point of view of their apparent morphology, HEGEWALD & SCHNEPF (1974) suggested the need to study populations to confirm the real existence of the two species.

Scenedesmus granulatus West & West is typical for the formation of disciform cenobia, whose broadly oval or spherical cells are arranged in two irregular rows, sometimes in different planes. In addition, the cell wall of the latter species shows wart-like granules, which are not structures of the sporopollenin layer of the wall, but are impregnated with inorganic substances (HEGEWALD *ET AL.* 1988).

Representatives of this species were found in two municipalities in the state of Sao Paulo, in both of which they only formed populations of a few individuals. With regard to morphological variation within this species, it is important to mention that a widely variable character was the presence or absence of wart-like granulations in the cell wall. *Scenedesmus verrucosus* Roll was not considered easy to identify taxonomically due to the fact that the latter character is extremely variable within populations, making it extremely difficult to identify representatives of the species.

Enallax Pascher 1943

Cenobium made up of (2)-4-8 elongated cells, more or less parallel to each other, slightly inclined towards each other (or crossed). The cells can be from fusiform to ellipsoidal, rarely somewhat asymmetrical, with the lateral margins being more or less convex, somewhat asymmetrical. Firm cell wall, with meridional ribs or striations. Single chloroplast per cell, parietal, with some thickening and a pyrenoid. Reproduction takes place by 2-4-8 autospores which group together to form the cenobium while still inside the mother cell, forming a system of folds in the cell wall. The release of the daughter cenobia occurs by the rupture of the maternal cell wall (COMAS (1996).

Only one species identified:

Enallax acutiformis (Schroder) Hindák (Fig. 140)

Studies on the Chlorococcal Algae (Chlorophyceae) 5: 37, pl. 8, fig. 1-2. 1990.

Basionimo: *Scenedesmus acutiformis* Schróder, Forschungsberichte aus der Biologischen Station zu Pión 5: 45, pl. 2, fig. 4a-b. 1897.

Cenobium flat, formed by 4 cells; cells arranged linearly, fusiform, poles markedly acuminate, outer and inner cells practically straight; cell wall with 1 fold in the cell wall; cell length 15.6-21.3 pm, width 2.6-5.2 pm; parietal chloroplastid, 1 pyrenoid.

Geographical distribution in the state of Sao Paulo

IN LITERATURE: nothing. First mention of the occurrence of the genus in Brazil.

EXAMINED MATERIAL: **Municipality of Pirassununga**, SP-225, km 23, Cascalho neighborhood, lake, ?-XII-1973, *P.A.C. Senna* (SP123900).

Comment

Enallax acutiformis (Schróder) Hindák is the only species in the Scenedesmoideae subfamily to have a rib-like structure in its cell wall. However, it is not a true rib but a fold, i.e. a crease in the cell wall. In addition, the poles of the cells seem to end in a thorn, which is also not a true thorn but a markedly acuminate cell pole.

According to FOTT & KOMÁREK (1983), *Enallax* comprises two species, E. *alpina* Pascher and E. *coelastroides* (Bohlin) Skuja, which are only distinguishable under typical conditions due to the existence of many intermediate morphotypes. Because of this polymorphism, some authors prefer to combine the two species into a single one (see KALINA & PUNCOCHAROVA 1977). This procedure was adopted by HINDÁK (1990) who, in addition to the two species mentioned above, also added E. *costatus* (Schmidle) Pascher. In this case, the correct name for the resulting species is E. *costatus* (Schmidle) Pascher. KALINA (1966) and KALINA & PUNCOCHAROVA (1977) considered E. *costatus* (Schmidle) Pascher to be a *Scenedesmus* species on the basis of data from transmission electron microscopy, i.e. because the cell wall consists of two layers, the inner of which is made up of cellulose and the outer of sporopollenin. The authors also stated that during the formation of autospores, plaques of sporopollenin appear between the daughter cells.

Enallax acutiformis Schróder was recorded in only one municipality in the state of São Paulo,

Pirassununga. Furthermore, only a few specimens of this species were found in a single population. However, they are very characteristic of the species. As for morphological variation, the only variable character was cell size. All the other characters remained stable within the single population analyzed for this species. We do not consider the specimens of *E. acutiformis* (Schroder) Hindák to be easy to identify because the cells appear to end in a thorn and can therefore be confused with representatives of some species of *Desmodesmus*.

Subfamily Tetrallantoideae *sensu* Komárek & Fott 1983

Different types of cenobium occur in this subfamily, in which the more or less elongated, ellipsoid, lunate, ovate or cylindrical cells, generally curved, are spatially oriented, showing a certain polarity, joining at one of their ends.

Tetrallantos Teiling 1916

Half-moon-shaped cells arranged in groups of four, sometimes forming syncenobia of 8-16 cells. Cenobia have a characteristic shape: two cells of varying curvature are arranged in the same plane, with the ventral face (concave) of one of them facing the same face as the other cell, touching at both poles. The other two cells of the cenobium are located in another plane (perpendicular to the previous one), each touching the joined poles of the first pair. Vegetative reproduction is carried out by autospores produced in numbers of two, four or eight per cell. When the autospores are released, they form a new cenobium. Each cell contains a parietal chloroplastid with a pyrenoid (BICUDO & MENEZES 2006).

Only one species identified:

***Tetrallantos lagerheimii* Teiling** (Fig. 156-158) SvenskBotanisk Tidskrift 10: 62. 1916.

Cenobium formed by 4 cells, 2 of which are in the same plane joined by both apices and the others are arranged vertically, touching their apices; lunate or roughly reniform cells; cell length 4.2-20.5 pm, width 2.2-6.7 pm; parietal chloroplastidium, 1 pyrenoid.

Geographical distribution in the State of Sao Paulo

IN LITERATURE: **Municipalities of Atibaia, Avaí, Juquiá, Pindamonhangaba, Santo André, Sao Paulo, Sumaré** and **Ubatuba** (SANT'ANNA 1984: 255, fig. 170-171). **Municipality of Juquiá** (SANT'ANNA *ET AL.* 1988: 91, fig. 46). **Municipality of Luiz Antonio** (SCHWARZBOLD 1992: 112, fig. 7). **Municipality of Ribeirao Preto** (SILVA 1999: 291, fig. 72). **Municipality of Sao Carlos** (HINO & TUNDISI 1977: 62, fig. 53). **Municipality of Sao Paulo** (SANT'ANNAETAL. 1989: 97, fig. 98; *FERRAGUTETAL.* 2005: 156, fig. 100).

EXAMINED MATERIAL: **Municipality of Capivari**, SP-3O8, km 132, pond, with *Typha, Eichhornia* and *Pistia,* 2O-III-199O, *A.A.J de Castro & C.E.M. Bicudo* (SP239O44). **Municipality of Guaratinguetá**, Clube dos 5OO, lake, O1-IV-1966, *C.E.M. Bicudo* (SP96965). **Municipality of Jundiai**, SP-36O, km 68, stream, O5-V-2OO, C.E.M. *Bicudo & S.P. Schetty* (SP365686). **Municipio de Sao Paulo**, Parque Estadual das Fontes do Ipiranga, Lago do IAG, O3-VIII-1998, *I.S. Vercellino,* 23°38'O8"S e 23°4O'18"S, 46°36'48"W e 46°38'OO"W (SP399779). **Municipio de Ubatuba**, no precise location indicated, 29-I-1966, *O. Montes & R.R. Martins* (SP9689O).

Comment

Tetrallantos is, according to SANT'ANNA (1984), a monospecific genus very well characterized by the previously described arrangement of the cells of the tetracellular cenobium.

SANT'ANNA (1984) also mentioned that the cenobial mucilage can be so hyaline that it needs to be highlighted with methylene blue solution. ROSA & OLIVEIRA (199O) agreed with SANT'ANNA (1984) by mentioning that the morphology and arrangement of the cells in each group of four leaves no doubt as to the identification of this genus and, consequently, its single species.

The popular ones currently studied are fully identified with the specimens described in KOMÁREK & FOTT (1983), however, when compared with the illustration in TEILING (1916: fig. 62) some morphological differences are visible, such as the cells being cylindrical, slightly arched and sometimes twisted at the end. These characteristics of the Veracruz population in Mexico (COMAS *ETAL.* 2OO7) have also been observed in materials from Cuba (COMAS 1996). However, given their insignificance, these differences are of no taxonomic importance.

Tetrallantos lagerheimii Teiling has been recorded in five locations in the state of São Paulo and has always occurred in populations with large numbers of individuals. It is also a species that is easy to identify due to the fact that it presents a particular arrangement of its cells when forming the cenobium. As for morphological variation, it only showed variability in terms of cell dimensions, while the other characters remained fairly stable within the species.

Chapter 5
Excluded taxa

Some names in the specialized literature were excluded from this work because, in some cases, they did not include a description of the material identified; other times, because they did not include measurements and/or illustrations; still other times, this information was presented too succinctly; and because they were simply included in lists of the material identified for a given environment in almost all works on phytoplankton or periphyton ecology. In all these cases, however, there was a lack of means to re-identify the material. When in doubt, and not knowing what the author actually identified, we preferred not to consider these names in this work.

The list of these taxa is as follows:

Coronastrum
XAVIER *ET AL.* (1985): citation only.

Crucigenia
XAVIER *ET AL.* (1985): citation only.

Crucigeniella crucifera (Wolle) Komárek
BICUDO ET AL. (1999a): citation only.
TUCCI *ET AL.* (2006): citation only.

Crucigenia fenestrata (Schmidle) Schmidle
SCHWARZBOLD (1992): illustration only.
BICUDO ET AL. (1999a): citation only.
TUCCI ET AL. (2006): citation only.

Crucigenia quadrata Morren
BICUDO ET AL. (1999a): citation only.
TUCCIETAL (2006): citation only.

Crucigema rectangularis (Nágeli) Gay BORGE (1918): citation only.
KLEEREKOPER (1939): quote only.

Crucigeniella rectangularis (Nágeli) Komárek BICUDOETAL. (1999a): citation only.
Tuccietal (2006): citation only.

Crucigema tetrapedia (Kirchner) West & West KLEEREKOPER (1939): citation only.
BICUDO ET AL. (1999a): citation only.
Tucci et *AL.* (2006): citation only.

Crucigenia sp.
ROLLA *ETAL.* (1990): citation only.

Didymocystis inermis (Fott) Fott
BICUDO ET AL. (1999a): citation only.

Didymocystis planktonic Korsikov
BICUDO *ETAL.* (1999a): citation only.

Didymogenes anómala (G.M. Smith) Hindák BICUDOETAL. (1999a): citation only. Tucci et.*AL.* (2006): citation only.

Didymogenes palatina Schmidle BICUDO *ETAL.* (1999a): citation only. Tucci et *AL. (2006):* citation only.

Dimorphococcus
WHITE (1963): quote only.
XAVIER ET AL. (1985): citation only.

Dimorphococcus lunatus A. Braun BORGE (1918): citation only. TUNDISI & HINO (1977): citation only.

Scenedesmus
WHITE (1960): quote only.
WHITE (1961): quote only.
WHITE (1962): quote only.
BRANCO (1963): just quoting.
BRANCO (1964): just quoting.
PALMER (1961): just quoting.
POTEL (1964): just quoting.
CHAVES (1978): just quoting.
XAVIER *ETAL.* (1985): just illustrating.

Scenedesmus abundans (Kirchner) Chodat
BICUDO ETAL. (1999a): just quoting.

Scenedesmus acuminatus (Lagerheim) Chodat
    XAVIER ETAL. (1985): only illustrated.
    TUCCI ETAL. (2006): just quoting.
Scenedesmus acuminatus (Lagerheim) Chodat var. *bemardii* (G.M. Smith) Dedusenko
    BICUDO ETAL (1999a): just quoting.
    Tucci ETAL. (2006): just quoting.
Scenedesmus acuminatus (Lagerheim) Chodat var. *elongatus* G.M. Smith
    BICUDO ETAL (1999a): just quoting.
    TUCCI ETAL. (2006): just quoting.
Scenedesmus acuminatus (Lagerheim) Chodat f *maximus* Uherkovich
    BICUDO ET.IL. (1999a): just quoting.
    TUCCIETAL. (2006): just quoting.
Scenedesmus acutus Meyen
    HIÑO & TUNDISI (1977): just illustrating.
    BICUDoETAL. (1999a): just quoting.
    TUCCIEE.IL. (2006): just quoting.
Scenedesmus acutus Meyen f. *alternans* Hortobágyi
    BICUDO ET AL. (1999a): just quoting.
    TUCCIET.AL. (2006): just quoting.
Scenedesmus anomalus (G.M. Smith) Ahlstrom & Tiffany var. *acaudatus* Hortobágyi
    BICUDO ET AL. (1999a): just quoting.
Scenedesmus arcuatus Lemmermann
    BICUDO ET AL. (1999a): just quoting.
    TUCCIET.AL. (2006): just quoting.
Scenedesmus arcuatus Lemmermann f *spinosus* Hortobágyi & Németh BICUDO *ETAL.* (1999a): cite
    only.
    Tucci *ETAL.* (2006): just quote.
Scenedesmus bicaudatus Dedusenko
    BICUDO *ETAL.* (1999a): just quote.
    Tucci *ETAL.* (2006): just quote.
Scenedesmus bicellularis Chodat
    CARDOSO (1979): apenas ilustrarán.
Scenedesmus bijuga (Turpin) Kützing
    KLEEREKOPER (1939): quote only.
Scenedesmus bijugatus (Turpin) Kützing
    BORGE (1918): just quote.
Scenedesmus bijugatus (Turpin) Kützing var. *alternans* (Reinsch) Hansgirg BORGE (1918): only citarán.
Scenedesmus bijugus (Turpin) Kützing var. *bijugus* SANT'ANNAETAL. (1989): only illustrated.
    SCHWARZBOLD (1992): only illustrated.
    BICUDO ET AL. (1999a): just quote.
    GENTIL (2000): only illustrate.
    Tucci et.*AL.* (2006): just quote.
Scenedesmus bijugus (Turpin) Kützing var. *disciformis* (Chndat) Leite BICUDO LEAL. (1999a): only
    citarán.
    TUCCILLAL (2006): just quote.
Scenedesmus brasiliensis Bnhlin
    BORGE (1918): just quote.
    BICUDO & BICUDO (1967): only citarán.
    MARINHO (1994): only describe.
Scenedesmus brevispina (G.M. Smith) Chndat
    BICUDOLLAL (1999a): just quote.
    Tncci L7.1L. (2006): just quote.
Scenedesmus carinatus (Lemmermann) Chndat
    BICUDOATAL. (1999a): just quote.
    Tucci *ETAL.* (2006): just quoting.
Scenedesmus carinatus (Lemmermann) Chodat var. *bicaudatus* Hortobágyi
    BICUDO ET AL. (1999a): just quoting.

Tuccietal (2006): just quoting.
Scenedesmus denticulatus Lagerheim
    BICUDO *ETAL.* (1999a): just quoting.
    Tuccietal (2006): just quoting.
Scenedemus denticulatus Lagerheim var. *australis* Playfair
    BICUDO *ETAL.* (1999a): just quoting.
    Tucci *ETAL. (2006):* just quoting.
Scenedesmus ecomis (Ralfs) Chodat
    HINO & TUNDISI (1977): just illustrating.
    TUNDISI & HINO (1981): just quoting.
Scenedemus ellipsoideus Chodat
    BICUDO *ETAL.* (1999a): just quoting.
    Tucci *ET.AL. (2006):* just quoting.
Scenedesmus ellipticus Rope
    MARINI IO (1994): just quoting.
    SILVA (1999): just illustrating.
Scenedesmus granúlalas West & West
    SANT'ANNAíTTL (1989): just illustrating.
    BICUDOETML (1999a): just quoting.
    Tucci *ETAL.* (2006): just quoting.
Scenedesmus hystrix Lagerheim
    BICUDOETAL. (1999a): just quoting.
Scenedesmus intermedias Chodat
    BICUDOíTAL. (1999a): just quoting.
Scenedesmus microspina Chodat
    TUNDISI & HINO (1981): just quoting.
Scenedesmus naegelii Brébisson
    SANT'ANNA (1984): just illustrating.
Scenedesmus obliquus (Turpin) Kützing
    BORGE (1918): just quoting.
Scenedesmus obliquus (Turpin) Kutzing var. *dimorphus* (Turpin) Rabenhorst
    BORGE (1918): just quoting.
Scenedesmus opoliensis P. Richter
    BICUDO ET.4L. (1999a): just quoting.
    TUCCIET^A. (2006): just quoting.
Scenedesmus opoliensis P. Richter var. *danubialis* Hortobágyi
    SILVA (1999): just illustrating.
Scenedesmus ovalternus Chodat
    BICUDO E7.4L. (1999a): just quoting.
    Tuccietal (2006): just quoting.
Scenedesmus ovalternus Chodat var. *graevenitzii* (Bernard) Chodat
    BICUDO £7AL. (1999a): just quoting.
    Tucci *ET.AL. (2006):* just quoting.
Scenedesmuspolyglobulus Hortobágyi
    BICUDOET.*AL.* (1999a): just quoting.
Scenedesmus protuberans Fritsch
    SCHETTY (1998): just illustrating.
    BICUDO ET.*AL.* (1999a): just quoting.
    Tucci et.*AL. (2006):* just quoting.
Scenedesmus quadricauda (Turpin) Brébisson
    BORGE (1918): just quoting.
    HINO & TUNDISI (1977): just illustrating.
    TUNDISI & HINO (1981): just quoting.
    BICUDO ET.*AL.* (1999a): just quoting.
    Tucci et.*AL. (2006):* just quoting.
Scenedesmus quadricauda (Turpin) Brébisson var. *longispina* (Chodat) G.M. Smith f. *asymmetricus*
(Hortobágyi) Uherkovich

BICUDO ET AL. (1999a): just quoting.
Tucci et.*AL.* (2006): just quoting.
Scenedesmus semipulcher Hortobágy
BICUDO ET.*AL.* (1999a): just quoting.
Scenedesmus serratus (Corda) Bohlin
BICUDOETAL. (1999a): just quoting.
Scenedesmus spinosus Chodat
BICUDO ET AL. (1999a): just quoting.
Scenedesmus verrucosus Roll
BICUDO ET AL. (1999a): just quoting.
Tucci *ET.AL.* (2006): just quoting.
Scenedesmus sp.
KLEEREKOPER (1939): just quoting.
ROLLA *ET.AL.* (1990): just quoting.
PERES & SENNA (2000): just illustrating.
Tetrallantos
XAVIER *ETAL.* (1985): just quoting.
Tetrallantos lagerheimii Teiling
TUNDISI & HIÑO (1981): just quoting.
BICUDO ET AL. (1999a): just quoting.
PERES & SENNA (2000): just quoting.
TucciET.lL. (2006): just quoting.
Tetrallantos sp.
ROLLA *ET.AL.* (1990): just quoting.
Tetrastrum
XANTERETAL. (1985): just quoting.
Tetrastrum elegans Playfair
BICUDO *ET AL.* (1999a): just quoting.
Tetrastrum heteracanthum (Nordstedt) Chodat
BICUDO ET AL. (1999a): just quoting.
TucciET.lL. (2006): just quoting.
Tetrastrum peterfii Hortobágyi
SANT'ANNA *ETAL.* (1989): just illustrating.
BICUDO ET AL. (1999a): just quoting.
Tucci L7.1/. (2006): just quoting.
Tetrastrumpunctatum (Schmidle) Ahlstrom & Tiffany
BICUDO *ET AL.* (1999a): just quoting.
TUCCIET.IL. (2006): citation only.
Tetrastrum triangulare (Chodat) Komárek
BICUDOETAL. (1999a): just quoting.
Tucci /.7.1/. (2006): citation only.
Tetrastrum sp.
ROLLA *ET AL.* (1990): citation only.
PERES & SENNA (2000): citation only.
Westella
XAVSEXLETAL. (1985): just illustrating.
Westella botryoides (West & West) De-Wild.emann BICUDO *ETAL.* (1999a): citation only.
TUCCIETAL (2006): just quoting.

Illustrate

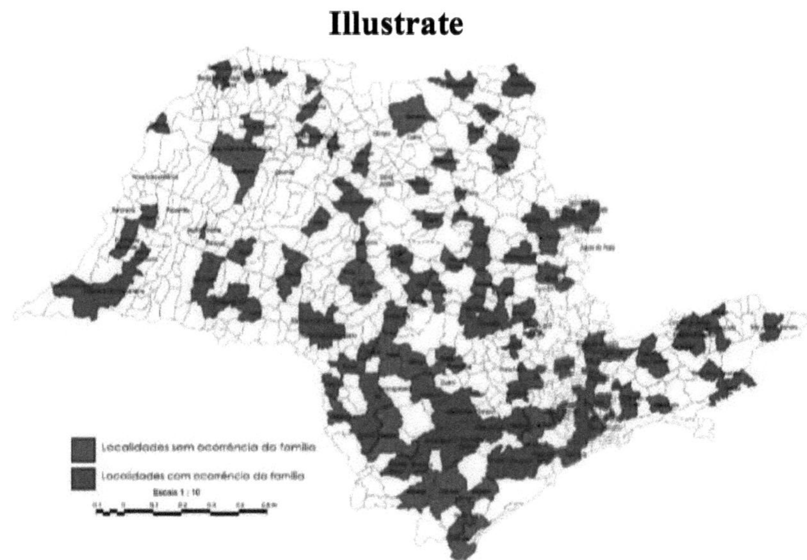

Figure 3: Localities sampled in the state of Sao Paulo and the distribution of the Scenedesmaceae family.

Fig. 4. Distribution of the Scenedesmaceae family in the state of Sao Paulo according to the literature consulted.

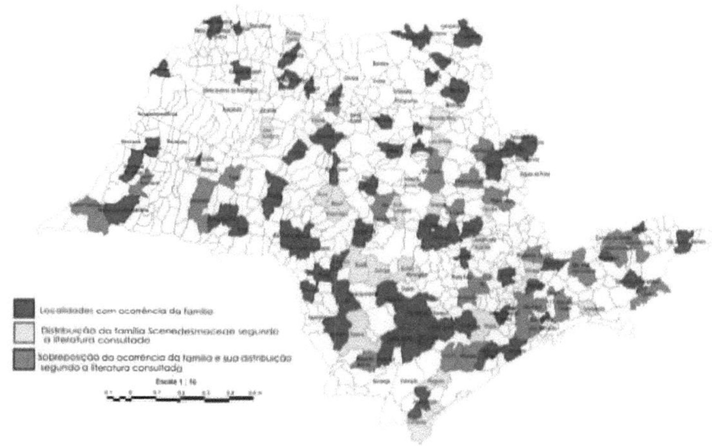

Figure 5 - Overlapping occurrence of the Scenedesmaceae family and its distribution according to the literature consulted.

Fig. 6.   *Coronastrum anglicum* Flint.
Fig. 7.   *Crucigenia fenestrata (*Schmidle) Schmidle (according to SANT'ANNA 1984).
Fig. 8.   *Crucigenia mucronata* (G.M. Smith) Komárek.
Fig. 9.   *Crucigenia quadrata* Morren.
Fig. 10.   *Crucigenia tetrapedia* (Kirchner) West & West.
Fig. 11.   *Crucigeniella apiculata* (Lemmermann) Komárek.
Fig. 12.   *Crucigeniella crucifera* (Wolle) Komárek.
Fig. 13.   *Crucigeniella rectangularis* (Nägeli) Komárek (according to FERRAGUT *ET AL.* 2005).
Fig. 14.   *Didymogenes anómala* (G.M. Smith) Hindák (according to SANT'*AnnaETAL.* 1989).
Fig. 15.   *Didymogenespalatina* Schmidle (according to KOMÁREK &FOTT 1983).
Fig. 16.   *Tetrachlorellaalternans* (G.M. Smith) Korsikov.
Fig. 17.   *Tetrastrum elegans* Playfair (according to SANT'ANNA 1984).
Fig. 18-20. *Tetrastrum heteracanthum* (Nordstedt) Chodat.
Fig. 21.   *Tetrastrum komarekii* Hindák (according to TUCCI *ETAL.* 2006).
NOTE: the scales of the figures are worth 10 pm, unless specifically indicated.

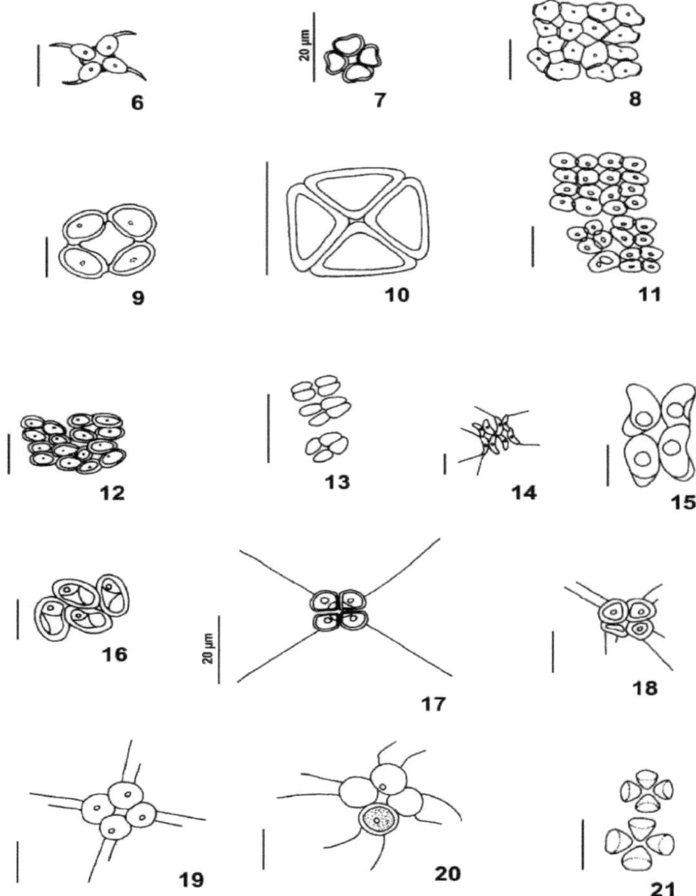

Fig. 22. *Tetrastrum staurogeniaeforme* (Schröder) Lemmermann (according to SANT'ANNA 1984).
Fig. 23. *Tetrastrum trianguläre* (Chodat) Komarek.
Fig. 24-26. *Westella botryoides* (W. West) De-Wildemann.
Fig. 27. *Willea irregularis* (Wille) Schmidle (according to SANT'ANNA 1984).
Fig. 28. *Willea vilhelmii* (Fott) Komarek.
Fig. 29. *Pseudotetrastrum punctatum* (Schmidle) Hindak.
Fig. 30-31. *Tetranephris brasiliensis* Leite & C. Bicudo.
Fig. 32. *Dimorphococcus lunatus* A. Braun.
NOTE: the scales of the figures are worth 10 pm. unless specifically indicated.

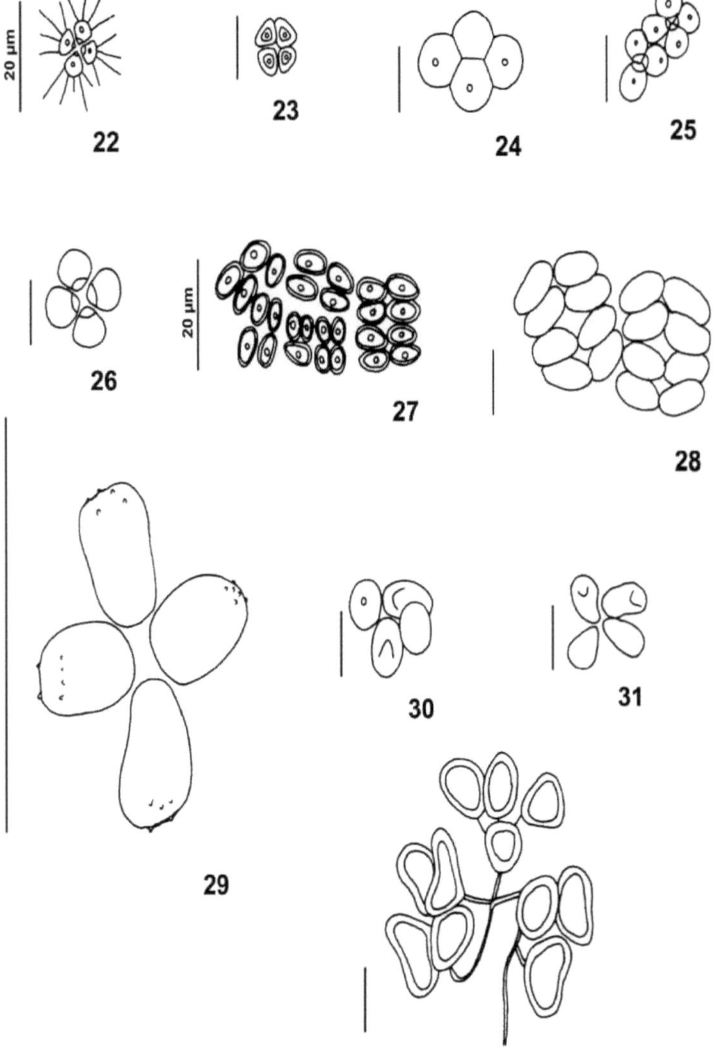

20 µm
22
23
24
25
20 µm
26
27
28
29
30
31
32

Fig. 33-38. *Desmodesmus abundans* (Kirchner) Chodat.

Fig. 39-41. *Scenedesmus aculeolatus* Reinsch

Fig. 42-46. *Desmodesmus armatus* (Chodat) Hegewald var. *armatus*.

Fig. 47-50. *Desmodesmus armatus* (Chodat) Hegewald var. *bicaudatus* (Guglielmetti) Hegewald.

Fig. 51-52. *Desmodesmus armatus* (Chodat) Hegewald var. *spinosus* (Fritsch & Rich) Hegewald.

Fig. 53-55. *Desmodesmus arthrodesmiformis* (Schröder) An, Friedl & Hegewald.

Fig. 56-58. *Desmodesmus brasiliensis* (Bohlin) Hegewald.

NOTE: the scales of the figures are worth 10 pm. unless specifically indicated.

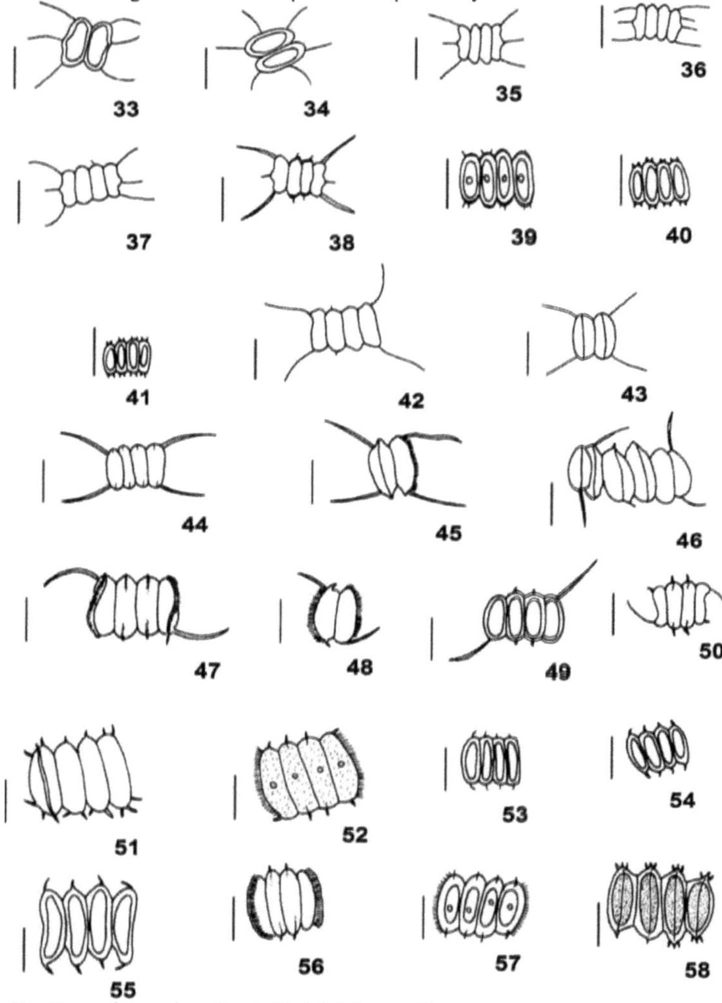

Fig. 59 - *Desmodesmus brasiliensis* (Bohlin) Hegewald.

Fig. 60-63. *Desmodesmus communis* (Hegewald) Hegewald.

Fig. 64 *Desmodesmus denticulatus* (Lagerheim) An, Friedl & Hegewald var. *denticulatus*.

Fig. 65-68. *Desmodesmus denticulatus* (Lagerheim) An, Friedl & Hegewald var. *linearis* (Hansgirg) Hegewald.

Fig. 69-70. *Desmodesmus dispar* Brébisson.

Fig. 71-72. *Desmodesmus flavescens* (R. Chodat) Hegewald var. *breviaculeatus* (Bourrelly)

Hegewald.

Fig. 73-74. *Scenedesmus gutwinskii* Chodat var. *bekesensis* Uherkovich

Fig. 75 - *Desmodesmus intermedias* (R. Chodat) Hegewald var. *intermedias* (according to FERRAGUT *ET AL.* 2005).

Fig. 76-81. *Desmodesmus intermedius* (R. Chodat) Hegewald var. *acutispinus* (Roll) Hegewald.

Fig. 82 *Desmodesmus lunatus* (West & West) Hegewald.

NOTE: the scales of the figures are worth 10 pm. unless specifically indicated.

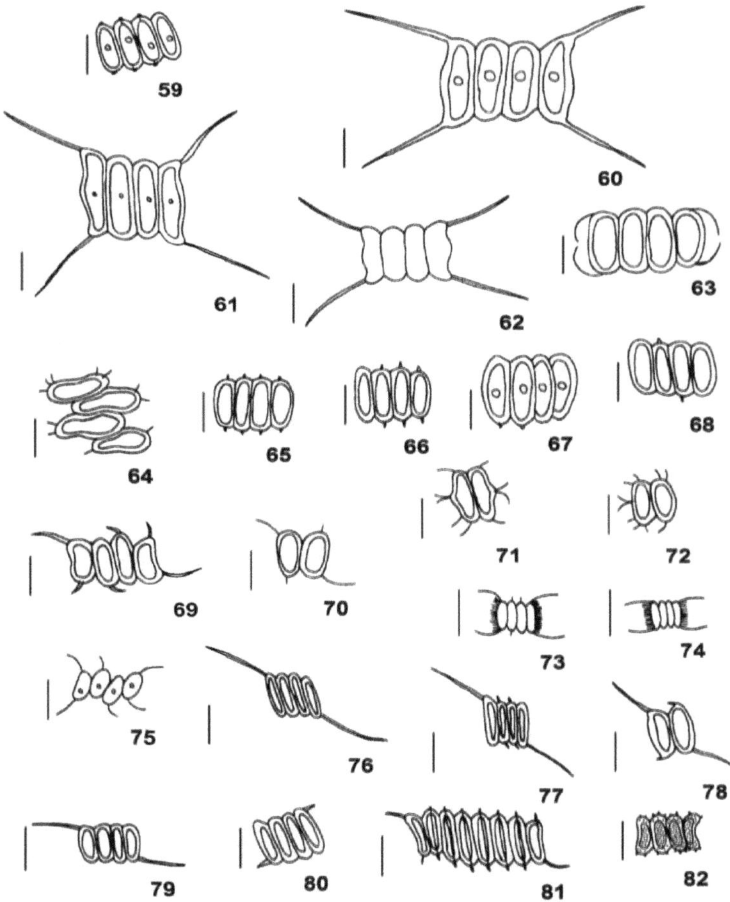

Fig. 83-84. *Desmodesmus lunatus* (West & West) Hegewald.

Fig. 85-87. *Desmodesmus maximus* (West & West) Hegewald.

Fig. 88 - *Desmodesmus opoliensis* (P. Richter) Hegewald var. *opoliensis* (according to SANT'ANNA 1984).

Fig. 89-91. *Desmodesmus opoliensis* (P. Richter) Hegewald var. *carinatus* (Lemmermann) Hegewald.

Fig. 92-94. *Desmodesmus opoliensis* (P. Richter) Hegewald var. *mononensis* (R. Chodat) Hegewald.

Fig. 95 - *Desmodesmusperforatus* (Lemmermann) Hegewald (according to SANT'ANNA 1984).

Fig. 96-97. *Desmodesmus pleiomorphus* (Hindak) Hegewald.

NOTE: the scales of the figures are worth 10 pm. unless specifically indicated.

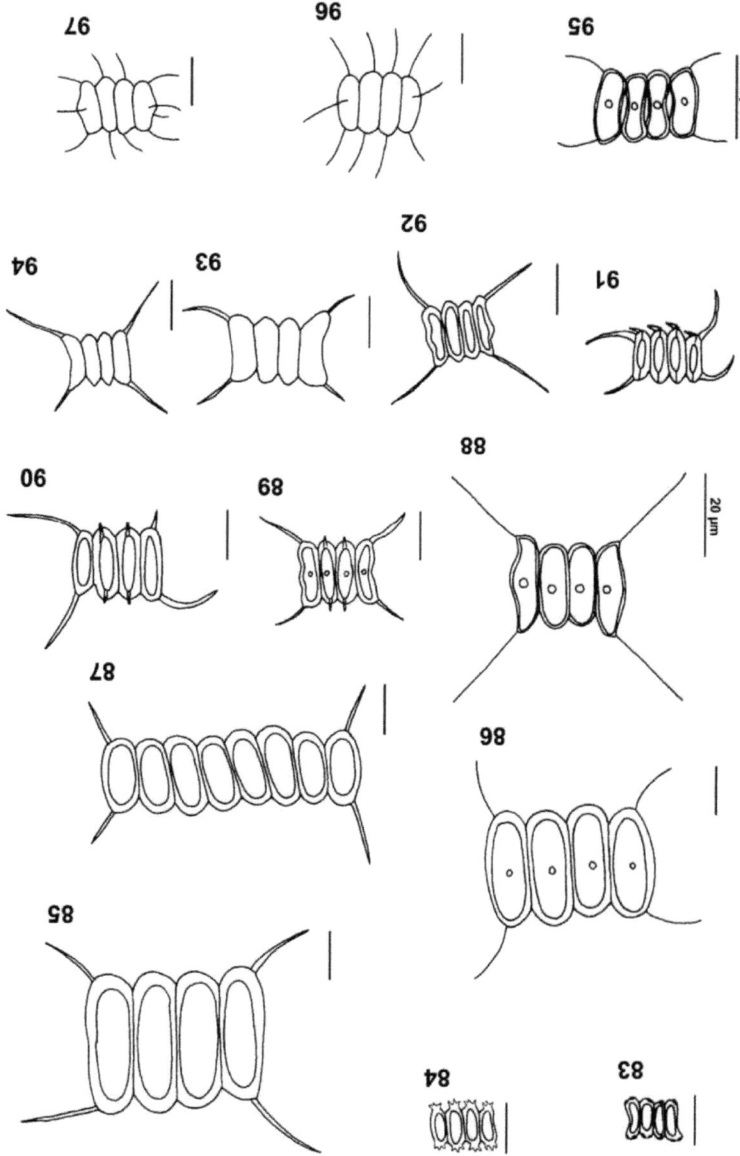

Fig. 98 - *Desmodesmusprotuberans* (Fritsch & Rich) Hegewald (according to SANT'ANNA 1984).
Fig. 99-100. *Desmodesmuspseudodenticulatus* (Hegewald) Hegewald.
Fig. 101-104. *Desmodesmus serratus* (Corda) An, Friedl & Hegewald.
Fig. 105-106. *Desmodesmus spinosus* (R. Chodat) Hegewald (according to FERRAGUT *ET AL*. 2005).
Fig. 107-109. *Desmodesmus spinulatus* (Biswas) Hegewald.
Fig. 110-111. *Desmodesmus subspicatus* (R. Chodat) Hegewald & Schmidt.
Fig. 112.   *Desmodesmus* sp. 1.
Fig. 113.   *Desmodesmus* sp. 2.
Fig. 114-115. *Desmodesmus* sp. 3.
Fig. 116. *Pseudodidymocystis fina* (Komárek) Hegewald & Deason (according to FERRAGUT *ET AL*. 2005).
Fig. 117. *Pseudodidymocystis planctonica* (Korsikov) Hegewald & Deason (according to TUCCI *ET AL*.
         2006).
NOTE: the scales of the figures are worth 10 pm, unless specifically indicated.

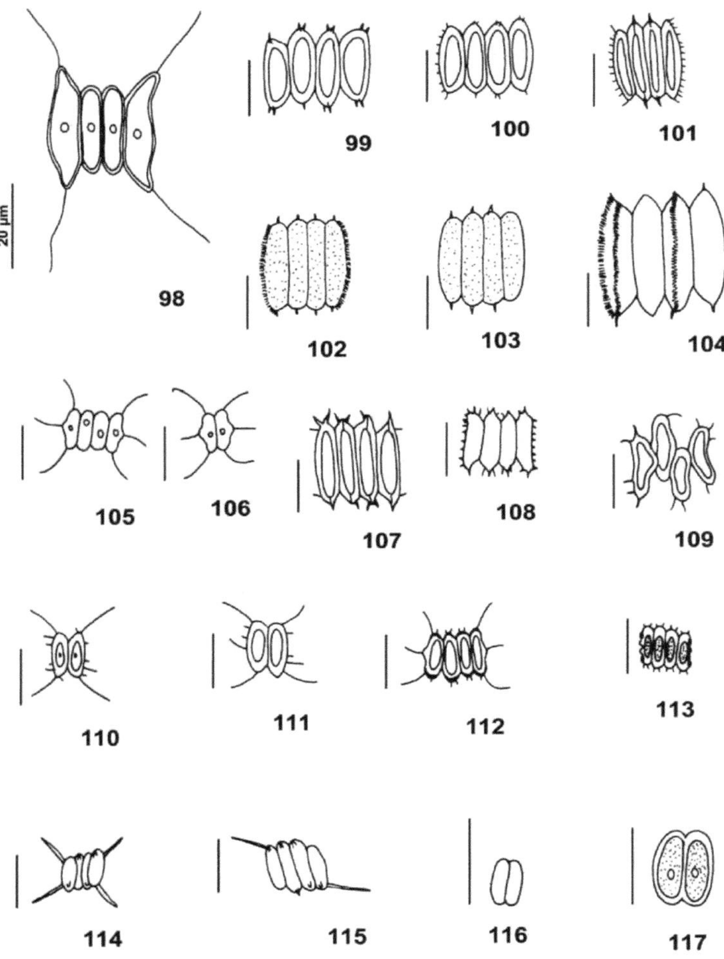

Fig. 118-119. *Scenedesmus acuminatus* (Lagerheim) Chodat var. *acuminatus.*
Fig. 120. *Scenedesmus acuminatus* (Lagerheim) Chodatvar. *elongatus* G.M. Smith (according to SANT'ANNA 1984).
Fig. 121-124. *Scenedesmus incrassatulus* Bohlin.
Fig. 125. *Scenedesmus indicus* Philipose *ex* Hegewald, Engelberg & Paschma (according to TUCCI *ETAL.* 2006).
Fig. 126. *Scenedesmusjavanensis* R. Chodat *var. javanensis.*
Fig. 127. *Scenedesmus javanensis* R. Chodat var. *schroeteri* (Huber-Pestalozi) Comas & Komarek (according to SANT'ANNA *ETAL.* 1989).
Fig. 128-137. *Scenedesmus obliquus* (Turpin) Kutzing var. *dimorphus* (Turpin) Kutzing.
Fig. 138. *Scenedesmus regularis* Svirenko (according to TUCCIETAL. 2006).
Fig. 139. *Scenedesmus wisconsinensis* (G.M. Smith) Chodat.
Fig. 140. *Enallax acutiformis* (Schroder) Hindak.
NOTE: the scales of the figures are worth 10 pm, unless specifically indicated.

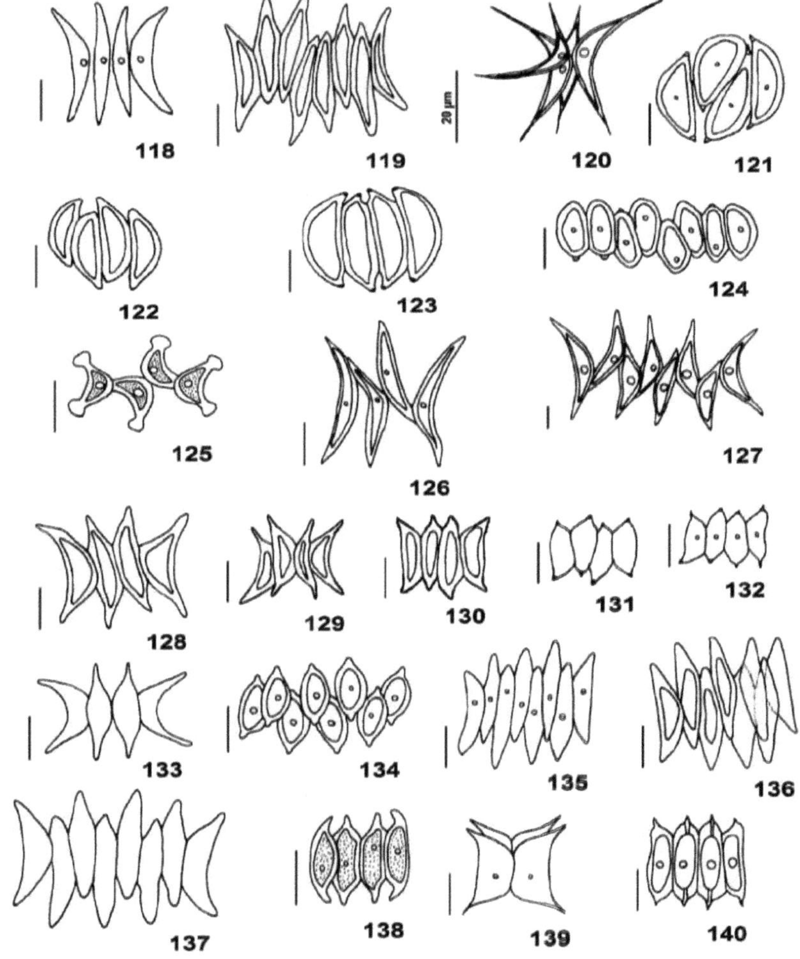

Fig. 141-143. *Scenedesmus acunae* Comas.
Fig. 144-145. *Scenedesmus arcuatus* (Lemmermann) Lemmermann var. *platydiscus* G.M.
    Smith.
Fig. 146-147. *Scenedesmus bicellularis* (Chodat) Komarek.
Fig. 148.   *Scenedesmus curvatus* Bohlin (according to SANT'ANNA 1984).
Fig. 149.   *Scenedesmus ecornis* (Ehrenberg) Chodat (according to FERRAGUT *ET AL.* 2005).
Fig. 150.   *Scenedesmus ellipticus* Corda (according to FERRAGUT *ET AL.* 2005).
Fig. 151-154. *Scenedesmus obtusus* Meyen.
Fig. 155 - *Scenedesmus verrucosus* Roll.
Fig. 156-158. *Tetrallantos lagerheimii* Teiling.
NOTE: the scales of the figures are worth 10 pm. unless specifically indicated.

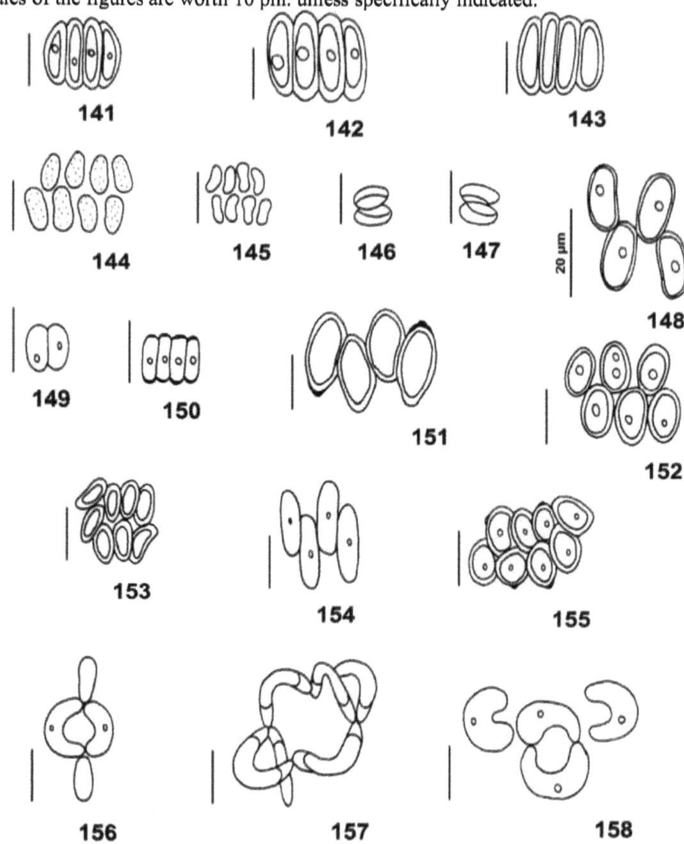

Conclusions and suggestions

The floristic survey of the Scenedesmaceae family in the area of the State of Sao Paulo, carried out by studying 153 sampling units, led to the following conclusions:

1. 73 Scenedesmaceae taxa were identified, distributed in 66 species (including three possible new species), 12 varieties that are not typical of their respective species and two new combinations, increasing knowledge of the family in the state by 29% and the geographical distribution of its representatives in the same area by 40%. Included among the 73 taxa above are 17 which are exclusive to the literature, i.e. which were not found again during the research.

2. Of the 73 taxa identified, the following nine are pioneering for the state of Sao Paulo and for Brazil: *Crucigeniella apiculata* (Lemmermann) Komárek, *Tetrachlorella alternans* (G.M. Smith) Korsikov, *Willea vilhelmii* (Fott) Komárek, *Desmodesmus arthrodesmiformis* (Schroder) An, Friedl & Hegewald, Desmodesmus *lunatus* (Chodat) Hegewald, *Desmodesmus pleiomorphus* (Hindák) Hegewald, *Enallax acutiformis* (Schroder) Hindák, *Scenedesmus aculeolatus* Reinsch and *Scenedesmus gutwinskii* Chodat var. *bekesensis* Uherkovich.

3. Of the 153 sampling units examined from 122 municipalities in the state of São Paulo, the presence of representatives of Scenedesmaceae was documented in 91 of them. In certain locations, specimens of several species were observed, for example in the municipality of Sao Paulo, where representatives of 58 species were identified, and in the municipalities of Pirassununga and Pindamonhangaba, where representatives of 15 and 22 species were identified respectively.

4. The species best represented in terms of number of infraspecific taxa in the Sao Paulo State area were: *Desmodesmus armatus* (Chodat) Hegewald with two varieties in addition to the species typical, *D. denticulatus* (Lagerheim) An, Friedl & Hegewald with one variety, *D. intermedius* (Chodat) Hegewald with one variety, *D. opoliensis* (P. Richter) Hegewald with two varieties, *Scenedesmus acuminatus* (Lagerheim) Chodat with one variety and .*S' javanensis* Chodat with one variety in addition to that typical of the species.

5. Considering the 122 municipalities currently floristically surveyed, the most widely distributed taxa in the São Paulo State area were: *Scenedesmus obliquus* (Turpin) Kützing var. *dimorphus* (Turpin) Hansgirg (in 22 municipalities), *Desmodesmus brasiliensis* (Bohlin) Hegewald (in 17 municipalities) and *D. armatus (*Chodat) Hegewald var. armatus, *D. communis (*Turpin) Hegewald and S *acunae* Comas (in 14 municipalities each).

6. The characters used to identify the species and taxonomic varieties of Scenedesmaceae are entirely morphological and related to the vegetative phase of their life history. These characters included: (/) the shape of the cell; (2) the size of the cenobium and the cell, the latter identified by its maximum length and width (both without considering the spines); *(3) the* total length of the spines; and (4) the type of cell wall decoration.

7. No sexual reproduction was observed in the materials studied, a fact that agrees with the worldwide literature which states that sexual reproduction does not occur among representatives of the Scenedesmaceae family.

8. Some phases of vegetative reproduction by cell division were observed, which the literature interprets as the mitotic type.

9. The analysis of the material collected in the state of Sao Paulo showed how frequent morphological variation is in populations of Scenedesmaceae and how much this polymorphism interferes with the taxonomic identification of the material. Major changes were seen in the shape of the margin of the outer cells of the cenobium, the size and orientation of the polar spines and the length and width of the cells; and smaller changes in the general shape of the cells, the number of cells per cenobium and the presence/absence of pyrenoids would make it extremely difficult to identify species and taxonomic varieties of Scenedesmaceae if the variation in diacritical and meristic characteristics were not known, but especially in diagnostic characteristics at the population level and their significance assessed depending on the amount of variation.

10. Due to the frequent occurrence of polymorphism, individuals found only a few times and/or in small numbers each time, despite the numerous preparations examined, were only identified when they showed well-defined diagnostic features and/or a morphological variation considered negligible, which allowed them to be unequivocally identified taxonomically. Some individuals, however, which together did not amount to 1% of all the specimens studied, were excluded because their taxonomic identification required either the observation or the confirmation of certain morphological features considered diagnostic, but which could not be detected in such specimens.

These individuals will be the subject of further studies in the future.

11. The taxa in which the greatest occurrence of polymorphism was found were the following: *Desmodesmus brasiliensis* (Bohlin) Hegewald, *D. armatus* (Chodat) Hegewald var. *armatus, D. abundans* (Kirchner) Chodat, *D. arthrodesmiformis* (Schroder) An, Friedl & Hegew and S. *obliquus* (Turpin) Kützing var. *dimorphus (Turpin)* Hansgirg.

12. The three maps of the state of São Paulo show, in the first, in red, the localities (municipalities) where specimens of Scenedesmaceae were found and, in blue, those where no representative of the family was found (Fig. 3). 3); in the second, in yellow, the distribution of the representatives of this family according to the literature (Fig. 4); and in the third, in red, orange and yellow, the overlap of the occurrence of Scenedesmaceae in the state considering the information in the literature and in the samples examined (Fig. 5).

The same floristic survey also made it possible to formulate the following suggestions for action and future work:

1. The great morphological and metric variability detected at the population level, together with the common occurrence of forms with abnormal or incomplete development, indicated, as has been pointed out before, the absolute need to examine population samples in the process of identifying species and taxonomic varieties of Scenedesmaceae. In addition, they suggested, as a matter of urgency, a taxonomic and nomenclatural revision of the family's representatives.

2. Morphological characteristics such as the linear size and projection of the spines, variation in the shape of the outer margin of the cenobium cells and the size of the cells vary greatly in Scenedesmaceae and can make it difficult to identify their species and infraspecific categories. This variability suggests the development of studies on the ecology of more and less variable species from the point of view of their morphology, in order to know the causes of this variability, their percentage of occurrence and, only then, to define their use in the taxonomy of the family and at which hierarchical taxonomic level to use them.

3. The separation of species and infra-specific categories on the basis of a few characteristics, especially just one, must be accepted with extreme caution and parsimony. For the time being, and until further studies are carried out, it is suggested that no species or taxonomic variety be proposed if such a proposition is based on only a few details.

4. To provide subsidies for ecology, genetics, cytology, physiology, biochemistry, molecular biology, etc. projects that require prior knowledge of the taxonomic composition of the local phycological flora.

5. Molecular biology is undoubtedly a very useful tool for studying the taxonomy, systematics and evolution of algae. If more fully exploited, this tool could show unequivocal results regarding the taxonomy and systematics of representatives of the Scenedesmaceae, as it already has. However, it is absolutely necessary to combine morphological expression with distinct genetic sequences.

References cited

Ahlstrom, E.H. & Tiffany, L.H. 1934. The algal genus *Tetrastrum*. American Journal of Botany 21: 499-507.

An, S.S., Friedl, T. & Hegewald, E. 1999. Phylogenetic relationships of *Scenedesmus* and *Scenedesmus-lke* coccoid green algae as referred from ITS-2 rDNA sequence comparisons. Plantbiology 1: 418-428.

Atkinson, A.W., Gunning, B.E.S. & John, P.L.C. 1972. Sporopollenin in the cell wall in *Chlorella* and other algae: ultrastructure, chemistry and incorporation of[14] C-acetate, studied in synchronous cultures. Planta 107: 1-32.

Barcelos, E.M. 2003. Evaluation of periphyton as a sensor of experimental oligotrophication in a eutrophic reservoir (Garças Lake, São Paulo). Master's dissertation, Universidade Estadual Paulista, Rio Claro, 118p.

Belkinova, D. & Mladenov, R. 2000. Veränderlichkeit der Zellmorphologie bei einigen Scenedesmus-Arten der Untergattung *Acutodesmus* (Chlorophyta, Chlorococcales). Algological Studies 96: 79-88.

Belkinova, D. & Mladenov, R. 2002. Morphologische Veränderlichkeit bei klonalen Culturen von *Scenedesmus nygaardii* Hub.-Pest. und *Scenedesmus bernardii* G.M. Smith (Chlorophyta). Algological Studies 104: 123-138.

Bernard, C. 1908. Protococcaceés et Desmidiacées d'eau douce recoltées à Java. Department d'Agriculture des Indes Néerlandaises, Batavia, 230 p.

Bicudo, C.E.M. & Bicudo, R.M.T. 1967. Floating communities of algae in an artificial pond in the parque do estado, SP, Brazil. Journal of Phycology 3: 233-234.

Bicudo, C.E.M. & Menezes, M. 2006. Genera of algae from Brazilian continental waters: key to identification and descriptions. RiMa Editora, Sâo Carlos, 489 p.

Bicudo, C.E.M. & Ventrice, M.R. 1968. Algae of the Lapa marsh, Itatiaia National Park, Brazil. Proceedings of the XIX Congresso Nacional Botánica, Sociedade Botánica do Brasil, Fortaleza, p. 3-30.

Bicudo, C.E.M., Bicudo, D.C., Castro, A.A.J. & Picelli-Vicentim, M.M. 1992. Phytoplankton of the dammed stretch of the Paranapanema River (Rosana Hydroelectric Plant), São Paulo State, Brazil. Revista Brasileira de Biologia 52: 293-310.

Bicudo, C.E.M., Ramírez R., J.J., Tucci, A. & Bicudo, D.C. 1999. Phytoplankton population dynamics in a eutrophied environment: Garças Lake, Sao Paulo. *In:* Henry, R. (ed.) Reservoir ecology: structure, firming and social aspects. FUNDBIO/FAPESP, Botucatu, p. 449-508.

Biesemeyer, K.F. 2005. Nictemeral variation in the structure and dynamics of the phytoplankton community as a function of water temperature during the dry and rainy seasons in a shallow mesotrophic urban reservoir (Lago das Ninféias), Parque Estadual das Fontes do Ipiranga. Master's dissertation, Instituto de Botánica, Sao Paulo, 153 p.

Bohlin, K. 1897. Die Algen der ersten Regnell'schen Expedition, 1 : Protococcoideen. Bihang till SvenskaVetenskapsakademie Handlingar 23: 7: 3-47.

Bold, H.C. & Wynne, M.J. 1978. Introduction to the algae: structure and reproduction. Prentice Hall Inc., New Jersey, 706 p.

Borge, O. 1918. Die von Dr. A. Lofgren in Sao Paulo gessammelten Süsswasseralgen. Arkiv for Botanik 15: 1-108.

Bourrelly, P. 1962. Quelques Chlorophycées des eaux douces françaises rares ou nouvelles. Biologischjaarboek Dodonaeae 30: 305-312.

Bourrelly, P. 1990. Les algues d'eau douce: initiation à la systématique, 1: les algues vertes. Éditions N. Boubée & Cie, Paris, Vol. 1, 572 p.

Branco, S.M. 1959. Some aspects of Hydrobiology important for Sanitary Engineering. RevistaD.A.E. 33-34: 1-24.

Branco, S.M. 1961a. Data on the variation in plankton level caused by changes in temperature and water viscosity. Revista D.A.E. 38: 43.

Branco, S.M. 1961b. Biology of the Alto Cotia reservoirs, 2: influence of the chemical factors of the waters on the qualitative alterations of the algological flora. Revista D.A.E. 42: 1-2.

Branco, S.M. 1962. Preventive and corrective control of algae in drinking water. RevistaD.A.E. 45: 60-75.

Branco, S.M. 1964. Henri Charles Potel and the biology of the waters of Sao Paulo. Revista D.A.E. 52:26-28.

Branco, S.M., Branco, W.C., Lima, H.A.S. & Martins, M.T. 1963. Identification and importance of the main genera of algae of interest for water and sewage treatment. Revista D.A.E. 48-50: 1-60.

Brunnthaler, J. 1915. Protococcales. *In:* Pascher, A. (ed.). Die Süsswasserflora von Deutschlands, Österreichs und der Schweiz, 5: Chlorophyceae 2. p. 52-205.

Buchheim, M.A., Michalopulos, E.A. & Buchheim, J.A. 2001. Phylogeny of the Chlorophyceae with special reference to the Sphaeropleales: a study of 18S and 26S rDNA data. Journal of Phycology 37: 819-835.

Cain, J.R. & Trainor, F.R. 1976. Regulation of gametogenesis in *Scenedesmus obliquus* (Chlorophyceae). Journal of Phycology 12: 383-390.

Cardoso, M.B. 1979. Phytophthora of the stabilization pond of Sao José dos Campos, State of Sao Paulo,

Brazil, excluding Bacillariophyceae. Master's dissertation, University of Sao Paulo, Sao Paulo, 230 p.

Cerione, E.M., Cavagioni, M.G., Breier, T.B., Barrela, W. & Almeida, V.P. 2008. Survey of planktonic algae species and water analysis of the Quinzinho de Barros Zoo lake, Sorocaba, Sao Paulo. RevistaEletrónica de Biologia 1: 18-27.

Chaves, C.M. 1978. Caracterização ecológica da autodepepuração de lagos do Parque Zoológico de Sao Paulo, Master's dissertation, University of Sao Paulo, Sao Paulo, 61 p.

Chodat, R. 1909. Étude critique et experimentale sur le polymorphisme des algues. Librairie Geoeg et Cie., Genève, 165 p.

Chodat, R. 1913. Monographie d'algues en culture pure. Matériaux pour la Flore Cryptogamique Suisse 4: 1-266.

Chodat, R. 1926. *Scenedesmus:* étude de génétique, de systématique expérimentale et d'hydrobiologie. Zeitschrif für Hydrologie 3: 71-258.

Comas, A. 1980. New and interesting Chlorococcales (Chlorophyceae) from Cuba. Acta Botanica Cubana 2:1-18.

Comas, A. 1984. Chlorococcales (Chlorophyceae) de algunos acuatorios de Pinar del Rio, Cuba. Acta Botanica Cubana 17: 1-60.

Comas, A. 1991. Taxonomische Übersicht der zönobialen Chlorokokkalalgen von Kuba, 3: Fam. Scenedesmaceae. Algological Studies 61: 55-94.

Comas, A. 1996. Chlorococcales dulciacuícolas de Cuba. Biblioteca Phycologica 99: 1265.

Comas, A. & Komárek, J. 1984. Taxonomy and nomenclature of several species of *Scenedesmus* (Chlorellales). Algological Studies 35: 135-157.

Comas, A., Novelo, E. & Tavera, R. 2007. Coccal green algae (Chlorophyta) in shallow ponds in Veracruz, Mexico. Algological Studies 124: 29-69.

Comas, A. & Sánchez, P. 2008. Algunas consideraciones y sugerencias sobre "la crisis en la taxonomía tradicional" con especial referencia a las algas verdes unicelulares (cocales). Boletín Sociedad Española de Ficologia: Algas 39: 16-20.

Crossetti, L.O. 2002. Effects of experimental nutrient depletion on the phytoplankton community in a shallow eutrophic reservoir, Lago das Garças, São Paulo. Master's dissertation, University of São Paulo, Ribeirão Preto, 119 p.

Deason, T.R., Silva, P.C., Watanabe, S. & Floyd, G.L. 1991. Taxonomic status of the species of the green algal genus *Neochloris.* Plant Systematics and Evolution 177: 213219.

Edwall, G. 1896. Index of plants in the herbarium of the Geographical and Geological Commission of S. Paulo. Boletim da Comissâo Geograficade São Paulo 11: 51-215 (algae p. 185-190).

Echenique, R.O., Nunez-Avellaneda, M. & Duque, S.R. 2004. Chlorococcales de la AmazoníaColombiana, 1: Chlorellaceae y Scenedesmaceae. Caldasia26: 37-51.

Edelstein, T. & Prescott, G.W. 1964. *Raysiella,* a new genus of Oocystaceae (Chlorophyta) from Spring Lake, Michigan. Phycologia4: 121-125.

Ettl, H. 1980. Grundriß der allgemeinen Algologie. G. Fischer, Jena, 549 p.

Ettl, H. 1981. Die neue Klasse Chlamydophyceae, eine natürliche Gruppe der Grünalgen (Chlorophyta), 1 : Systematische Bemerkungen zun den Grünalgen. Plant Systematics and Evolution 137: 107-126.

Ettl, H, & Komárek, J. 1982. Was versteht man unten dem Begriff "coccale Grünalgen". Algological Studies 29: 345-374.

Ferragut, C., Lopes, M.R.M., Bicudo, D.C., Bicudo, C.E.M. & Vercellino, I.S. 2005. Periphytic and planktonic phycoflora (except Bacillariophyceae) of a shallow oligotrophic reservoir (Lago do IAG, Sâo Paulo). Hoehnea 32: 137-184.

Fermino, F.S. 2006. Seasonal evaluation of the effects of N and P enrichment on periphyton in a shallow mesotrophic tropical reservoir (Lago das Ninféias, Sâo Paulo). PhD thesis, Universidade Estadual Paulista, Rio Claro, 201p.

Ferreira, R.A.R. 2005. Structure of the periphytic algae community adhered to the aquatic macrophyte *Eichhornia azurea* Kunth in two lagoons located at the mouth of the Paranapanema River at the Jurumirim Dam. Doctoral thesis, University of São Paulo, São Carlos, 245 p.

Floyd, G.L., Watanabe, S. & Deason, T.R. 1993. Comparative ultrastructure of the zoospores of eight species of *Characium* (Chlorophyta). Archiv fur Protistenkunde 143: 63-73.

Fonseca, B.M. 2005. Phytoplankton diversity as an environmental discriminator in two shallow reservoirs with different trophic states in the Parque Estadual das Fontes do Ipiranga. PhD Thesis, University of Sâo Paulo, São Paulo, 208 p.

Fott, B. 1946. Taxonomical studies on Chlorococcales, 1. Studia botanica cechoslovaca 7: 165-171.

Fott, B. 1971. Algenkunde. G. Fischer, Jena, 581 p. (2ª edition).

Fott, B. 1973. Die Gattungen *Dicelulla* Swir., *Didymocystis* Korsch. und ihre Beziehungen zur Gattung *Scenedesmus* Meyen. Preslia 45: 1-10.

Fott, B. & Komárek, J. 1960. Das Phytoplankton der Teiche im Tschener Schlesien. Preslia 32:113-141.

Friedl, T. 1995. Inferring taxonomic positions and testing genus level assignments in coccoid green algae: a phylogenetic analysis of 18S ribosomal RNA sequences from *Dictyochloropsis reticulata* and from members of the genus *Myrmecia* (Chlorophyta, Trebouxiophyceae Cl. nov.). Phycologia 31: 632-639.

Friedl, T. & Rokitta, C. 1997. Species relationships in the lichen alga *Trebouxia* (Chlorophyta, Trebouxiophyceae): molecular phylogenetic anylises of nuclear-encoded large subunit rRNA gene sequences. Symbiosis 23: 125-148.

Gentil, R.C. 2000. Seasonal variation of the phytoplankton of a subtropical eutrophic lake and sanitary aspects, São Paulo, SP. Master's thesis, University of São Paulo, São Paulo, 198 p.

Gutwinski, R. 1890. Zur Wahrung der Priorität: Vorläufige Mittheilungen über einige neue Algen-Species und Varietäten aus der Umgebung von Lemberg. Botanische Centralblatt 43: 65-73.

Gutwinski, R. 1891. Flora glonow okolic Lwowa (Flora algarum agri leopoliensis). Sprawozdanie Komisyi Fizyogeograficznej Akademii Umiejetnosci w. Krakowie 27: 1124.

Hansgirg, A. 1888. Ueber die Süsswasserangen-Gattungen *Trochisia* Kütz. *(Acanthococcus* Lagerh., *Glochiococcus* De Toni) und *Tetraedron* Kütz. *(Astericium* Corda, *Polyedrium* Nag., *Ceraterias* Reinsch). Hediwigia27: 126-132.

Hegewald, E. 1977. *Scenedesmus communis* Hegewald, a new species and its relation to *Scenedesmus quadricauda* (Turp.) Breb. Algological Studies 18: 142-155.

Hegewald, E. 1978. Eine neue Unterteilung der Gattung *Scenedesmus* Meyen. Nova Hedwigia29: 343-376.

Hegewald, E. 1979. Vergleichende Beobachtungen an Herbarmaterial und Freilandmaterial von *Scenedesmus*. Algological Studies 24: 264-286.

Hegewald, E. 1985. Investigations on the lakes of Peru and their phytoplankton, 7: Algae of Laguna Yarinacocha, Pucallpa, with special reference to *Scenedesmus denticulatus* var. *linearis*. Algological Studies 40: 419-424.

Hegewald, E. 1988. A new interpretation of the genus *Didymocystis* Korsikov (Chlorophyta, Chlorococcales). Algological Studies 48: 309-312.

Hegewald, E. 1989. The *Scenedesmus* strains of the Culture Collection of the University of Texas at Austin (UTEX). Algological Studies 55: 153-189.

Hegewald, E. 1993. Studies on *Scenedesmus flavescens* Chod. (= S. *tenuispina* Chod. Algological Studies 71: 1-12.

Hegewald, E. 2000. New combinations in the genus *Desmodesmus* (Chlorophyceae, Scenedesmaceae). Algological Studies 96: 1-18.

Hegewald, E., An, S.S. & Tsarenko, P. 1998. Revision of *Scenedesmus intermedius* Chod. (Chlorophyta, Chlorococcales). Algological Studies 88: 67-104.

Hegewald, E., Coesel, P.F.M. & Hegewald, P. 2002. A phytoplankton collection from Bali, with the description of a new *Desmodesmus* species (Chlorophyta, Scenedesmaceae). Algological Studies 105: 51-78.

Hegewald, E. & Deason, T.R. 1989. *Pseudodidymocystis,* a new genus of Scenedesmaceae (Chlorophyceae). Algological Studies 55: 119-127.

Hegewald, E., Engelberg, K.E. & Paschma, R. 1988. Beitrag zur Taxonomie der Gattung *Scenedesmus,* Subgenus *Scenedesmus* (Chlorophyceae). Nova Hedwigia47: 497-533.

Hegewald, E. & Hanagata, N. 2000. Phylogenetic studies on Scenedesmaceae (Chlorophyta). Algological Studies 100: 29-49.

Hegewald, E., Jeeji-Bai, N. & Hesse, M. 1975. Taxonomische und floristische Studien nach Planktonalgen aus ungarischen Gewässern. Archiv fur Hydrobiologie, Supplement 46: 392-432.

Hegewald, E., Schmidt, A., Braband, A. & Tsarenko, P. 2005. Revision of the *Desmodesmus* (Sphaeropleales, Scenedesmaceae) species with lateral spines, 2: the multi-spinedto spineless taxa. Algological Studies 116: 1-38.

Hegewald, E., Schmidt, A. & Schnepf, E. 2001. Revision der lateral bestachelten Desmodesmus-Arten, 1: *Desmodesmus subspicatus* (R. Chodat) E. Hegewald & A. Schmidt. Algological Studies 101: 1-26.

Hegewald, E. & Schnepf, E. 1974. Beitrag zur Kenntnis der Grünalgenart *Scenedesmus verrucosus* Roll. Archiv für Hydrobiologie, Supplement 46: 151-162.

Hegewald, E., Schnepf, E. & Aldave, A. 1980. Investigations on the lakes of Perú and their phytoplankton, 5: the algae of Laguna Piuray and Laguna Huaypo, Cuzco. Algological Studies25: 387-420.

Hegewald, E. & Silva, P. 1988. Annotated catalogue of *Scenedesmus* and nomenclaturally related genera, including original descriptions and figures. Bibliotheca Phycologica 80: 1587.

Hegewald, E. & Wolf, M. 2003. Phylogenetic relationships of *Scenedesmus* and *Acutodesmus* (Chlorophyta, Chlorophyceae) as inferred from 18S rDNA and ITS-2 sequence comparisons. Plant Systematics and Evolution 241: 185-191.

Heynig, H. 1989. Interesting Phytoplankter aus Gewässern des Bezirks Halle (DDR), 6. Archiv für Protistenkunde 137: 57-68.

Heynig, H. & Krienitz, L. 1987. Interesting coccale Grünalgen (Chlorellales) aus einem Altwasser der Elbe (DDR). Archiv für Protistenkunde 134: 49-58.

Hindák, F. 1970. Culture collection of algae at Laboratory of Algology in Trebon. Algological Studies 2-3: 86-126.

Hindák, F. 1974. The Chlorococcal algal genus *Didymogenes* Schmidle 1905. Biológia 29: 559-570.
Hindák, F. 1977. Studies on the chlorococcal algae (Chlorophyceae), 1. Biologické Prace 23: 1-190.
Hindák, F. 1980. Studies on the chlorococcal algae (Chlorophyceae), 2. Biologické Prace 26: 1-195.
Hindák, F. 1984. Studies on the chlorococcal algae (Chlorophyceae), 3. Biologické Prace 30: 1-308.
Hindák, F. 1988. Studies on the chlorococcal algae (Chlorophyceae), 4. Biologické Prace 34: 1-263.
Hindák, F. 1990. Studies on the chlorococcal algae (Chlorophyceae), 5. Biologické Prace 36: 1-225.
Hino, K. & Tundisi, J. 1977. Atlas of Algae from the Broa Reservoir, 5. Federal University of Sao Carlos, Sao Carlos, 143 p.
Hoehne, F.C. 1948. Aquatic plants. Secretaria da Agricultura do Estado de Sao Paulo, Sao Paulo, 168 p. (Instituto de Botánica, série D).
Holtmann, T. 1994. Revision des Subgenus *Acutodesmus,* Gattung *Scenedesmus* (Grünalgen. Chlorophyceae). Doctoral Thesis, University of Essen, Fachbereich, Vol. 1-2, 542 p.
Holtmann, T. & Hegewald, E. 1986. Der Einfluß von Nährlösungen auf die Variabilität von Isolaten der Gattung *Scenedesmus* Untergattung *Acutodesmus.* Algological Studies 44: 365-380.
Hortobágyi, T. 1948. Üjab adatok a Balaton microvegetációjához. Dunántuli tudományos intézet kiadványai 10: 1-16.
Hortobágyi, T. 1967. Neue Beiträge zur Kenntnis der *Scenedesmus* ungars. Acta Botanica Hungarica 13:21-59.
Huber-Pestalozzi, G. 1929. Algologische Mitteilungen, 6. Archiv für Hydrobiologie 20: 413426.
Huszar, V.L.M. 1985. Planktonic algae of the Juturnaíba Lagoon, Araruama, RJ, Brazil. Revista Brasileira de Botánica 8: 1-19.
Joly, A.B. 1963. Genera of freshwater algae from the city of Sao Paulo and its surroundings. Rickia suppl. 1: 1-188.
Johnson, J.L., Fawley, M.W. & Fawley, K.P. 2007. The diversity of *Scenedesmus* and *Desmodesmus* (Chlorophyceae) in Itasca State Park, Minnesota, USA. Phycologia 46: 214-229.
Kalina, T. 1966. Morphologie und systematische Eingliederung der Art *Scenedesmus costatus* Schmidle (Chlorococcales). Preslia38: 346-350.
Kalina, T. & Puncocharova, M. 1977. Taxonomy and morphological comparison of three chlorococcal algae: *Scotiella oocystiformis* Lund., *Enallax coelastroides* (Bohl.) Skuja and *Scenedesmus costatus* Schmidle. Algological Studies 19: 105-141.
Kleerekoper, H. 1937. Biology of the Santo Amaro Old Dam (Guarapiranga Dam). Boletim R.A.E. 1: 151-161.
Kleerekoper, H. 1939. Limnological study of the Santo Amaro reservoir in São Paulo. Boletim da Faculdade de Filosofía, Ciéncias e Letras, Botánica series 2: 11-151.
Komárek, J. 1973. Taxonomische Begrenzung der Gattung *Didymocystis* Kors. (Scenedesmaceae, Chlorococcales). Preslia45: 311-314g.
Komárek, J. 1974. Taxonomische Bemerkungen zu einigen Arten der Mikroflora der Teiche in Böhmen. Acta Scientiarum Naturalium Musei Bohemici Meridionalis Ceské Budejovice 14: 161-190.
Komárek, J. 1975. New coenobial Chlorococcales of Cuba. Preslia 47: 275-279.
Komárek, J. 1983. Contribution to the chlorococcal algae of Cuba. NovaHedwigia 37: 65180.
Komárek, J. 1987. Species concept in coccal green algae. Algological Studies 45: 437-471.
Komárek, J. & Fott, B. 1983. Chlorophyceae (grünalgen) Ordnung: Chlorococcales. *In:* Huber-Pestalozzi, G. (org.) Das Phytoplankton des Süsswassers: Systematic und Biologie. E. Schweizerbart'sche Verlagsbuchhandling (Nägele u. Obermiller), Stuttgart, Vol. 7(1), 1044 p.
Komárek, J. & Ludvík, J. 1972. Die Zellwandstruktur als taxonomisches Merkmal in der Gattung *Scenedesmus,* 2: taxonomische Auswertung der untersuchten Arten. Algological Studies6: 11-47.
Komárek, J., Ludvík, J. & Pelicaric, S. 1977. Electronmicroscopical study of the cell wall of *Scenedesmus opoliensis.* Algological Studies 18: 33-45.
Komárková-Legnerová, J. 1969. The systematics and ontogenesis of the genera *Ankistrodesmus* Corda and *Monoraphidium* gen. nov. *In:* Fott, B. (ed.). Studies in Phycology, Stuttgart, p. 75-122.
Korsikov, O.A. 1953. Pidklas Protokokovi (Protococcineae). Viznacnik prisnovodnich vodorostej Ukrainskoj RSR 5: 1-439.
Krienitz, L. 1987. Studien zur Morphologie und Taxonomie der Untergattung *Acutodesmus* (Chlorellales). Algological Studies 46: 1-37.
Krienitz, L., Hegewald, E., Hepperle, D. & Wolf, M. 2003. The systematics of coccoid green algae: 18S rRNA gene sequence dataversus morphology- Biologia 58: 437-446.
Krienitz, L., Hegewald, E., Hepperle, D., Huss, V.A.R., Rohr, T. & Wolf, M. 2004. Phylogenetic relationship of *Chlorella* and *Parachlorella* gen. nov. (Chlorophyta, Trebouxiophyceae). Phycologia 43: 529-542.
Kützing, F.T. 1834. Synopsis diatomacearum oder Versuch einer systematischen Zusammenstellung der Diatomeen. Schwetschke, Halle, 92 p.
Leite, C.R. 1974. Contribution to the knowledge of the planktonic Chlorococcales (Chlorophyceae) of the Parque Estadual das Fontes do Ipiranga, Sao Paulo, Brazil. Master's thesis, University of Sao Paulo, Sao

Paulo, 151 p.

Leite, C.R. 1979. Chlorococcales (Chlorophyceae) from the State of Sao Paulo, Brazil. Doctoral dissertation, University of Sao Paulo, Sao Paulo, 410 p.

Leite, C.R. & Bicudo, C.E.M. 1977. *Tetranephris*, a new genus of Chlorococcales (Chlorophyceae) from southern Brazil. Phycologia 16: 231-233.

Lopes, M.R.M. 1999. Disturbance events affecting phytoplankton biomass, composition and species diversity in a shallow oligotrophic tropical lake (Lago do Instituto Astronómico e Geofísico, Sao Paulo, SP). PhD Thesis, University of Sao Paulo, Sao Paulo, 213 p.

Lukavsky, J. 1991. Motile cells in *Scenedesmus obliquus* in outdoor mass culture. Archiv für Protistenkunde 140: 345-348.

Marchand, L. 1895. Synopsis et tableau systématique des famillies qui composent la Classe des Phycophytes (Algues, Diatomées et Bakteriens).- *In:* Sousregne des Cryptogames, Société d'État Scientifique, Paris, 20 p. (2ª edition).

Marinho, M.M. 1994. Phytoplankton community dynamics of a small shallow reservoir densely colonized by submerged aquatic macrophytes (Jacaré reservoir, Mogi- Guaçu, SP, Brazil). Master's dissertation, University of Sao Paulo, Sao Paulo, 150 p.

Martins-da-Silva, R.C.V. 1996. New occurrences of Chlorophyceae (Algae, Chlorophyta) for the State of Pará. Boletim do Museu Paraense Emilio Goeldi, Botánica series 12(1): 1-16.

Mattox, K.R. & Stewart, K.D. 1984. Classification of the green algae: a concept based on comparative cytology. In: Irvine, D.E.G. & John, D.M. (eds.). The systematics of the green algae. Academic Press, London, p. 29-72.

Melkonian, M. 1990. Chlorophyte orders of uncertain affinities: order Microthamniales. In: Margulis, L., Corliss, J.O., Melkonian, M. & Chapman, D.J. (eds.). Handbook of Protoctista. Jones & Barlett Publishers, Boston, p. 652-654.

Melkonian, M. & Surek, B. 1995. Phylogeny of the Chlorophyta: congruence between ultrastructural and molecular evidence. Bulletin de la Société Zoologique de France 120: 191-208.

Meyen, F.J.F. 1829. Beobachtungen über einige niedere Algenformen. Verhandlungen der K. Leopoldinisch-carolinischen deutschen Akademie der Naturforscher 14: 768-778.

Mladenov, R. & Furnadzieva, S. 1999. Ontogenetische Veränderungen in klonalen Kulturen von *Scenedesmus acuminatus* (Lagerh.) Chod. und *Scenedesmus pectinatus* Meyen. Algological Studies 92: 35-46.

Moura, A.T. 1996. Structure and dynamics of the phytoplankton community in a eutrophic lagoon, Sao Paulo, SP, Brazil, at short time intervals: comparison between rainy and dry seasons. Master's dissertation, Universidade Estadual Paulista, Rio Claro, 172 p.

Näegeli, C. 1849. Gattungen einzelliger Algen: physiologisch und systematisch Bearbeitet. Friedrich Schulthess, Zürich, 133 p.

Nogueira, I.S. 1991. Chlorococcales *sensu latu* (Chlorophyceae) from the Municipality of Rio de Janeiro and surroundings: inventory and taxonomic considerations. PhD Thesis, University of São Paulo, São Paulo, 322 p.

Nordstedt, O. 1882. *In:* Wittrock, V. & Nordstedt, O. (eds). Algae aquae dulcis exsiccatae praecipue scandinavicae quas adjectis algis marinis chlorophyllaceis et phycochromaceis. O.L. Svanbäcks Boktryckeri Aktiebolac, Lundae, Fasc. 9-10.

O'Kelly, C.J. & Floyd, G.L. 1984. Correlations among patterns of sporangial structure and development, life histories, and ultrastructural features in the Ulvophyceae. In: Irvine, D.E.G. & John, D.M. (eds.). The systematics of the green algae, Academic Press, London, p. 121-156.

Ooshima, K. 1981. Taxonomic studies on *Scenedesmus* in Japan, 1: on *Scenedesmus acuminatus* (Lag.) Chod. and its varieties and S. *javanensis* Chod. Japanese Journal of Phycology 29: 85-93.

Palmer, C.M. 1961. Algae and water supplies in the São Paulo area. U.S. Department of Health, Education, and Welfare Technical Report. 7 p.

Pascher, A. 1915. Einzellige Chlorophyceen Gattungen unsicherer Stellung. In: Pascher, A. (ed.). Die Süsswasserflora Deutschlands, Österreichs und der Schweiz, Vol. 5, Chlorophyceae 2. Gustav Fischer, Jena, p. 206-236.

Peres, A.C. &. Senna, P.A.C. 2000. Chlorophyta of Diogo Lagoon. In: Santos, J.E. & Pires, J.S.R. (ed.). Jataí Ecological Station. RiMa Editora, São Carlos, vol. 2, p. 469481.

Philipose, M.T. 1967. Chlorococcales. Indian Council of Agricultural Research, New Delhi, 365 p.

Picelli-Vicentim, M.M. 1987. Planktonic Chlorococcales from the Iguaçu Regional Park, Curitiba, State of Paraná. Revista Brasileira de Biologia 47: 57-85.

Potel, H.C. 1964. Henri Charles Potel and the biology of the waters of São Paulo: observations on the chemical composition of the waters. Revista DAE 52: 26-28.

Printz, H. 1927. Chlorophyceae. In: Engler, A. & Prantl, K. (eds). Die natürlichen Pflanzenfamilien. Verlag vonWilhelm Engelmann, Leipzig, vol. 3, 463 p.

Ralfs, J. 1845. On the British Desmidieae. Annals and Magazine of Natural History 15: 401406.

Ramirez R., J.J. 1996. Vertical and nictemeral spatial variations in phytoplankton community structure and environmental variables on four sampling days at different times of the year in Garças Lake, São Paulo. PhD thesis, University of São Paulo, São Paulo, 283 p.

Reháková, H. 1969. Die Variabilität der Arten der Gattung *Oocystis* A. Braun. *In:* Fott, B. (ed.). Studies in Phycology, Stuttgart, p. 145-198.

Rogers, C.E., Mattox, K.R. & Stewart, K.D. 1980. The zoospore of *Chlorokybus atmosphyticus,* a Charophyceae with sarcinoid growth habit. American Journal of Botany 67: 774-783.

Rolla, M.E., Diabes, M.B.S., Franca, R.C. & Ferreira, E.M.V.M. 1990. Limnological aspects of the Volta Grande Reservoir, Minas Gerais/Sao Paulo. Acta Liminologica Brasiliensia 3: 236-247.

Roque, R. 1980. Ecological sanitary aspects and phytoplankton in Billings Reservoir. Master's thesis, University of Sao Paulo, Sao Paulo. 154 p.

Rosa, Z.M. & Oliveira, M.B. 1990. Chlorococcales (Chlorophyceae) from water bodies in the Municipality of Sao Jerónimo, Rio Grande do Sul, Brazil. Iheringia: Botánica series 40: 89114.

Sant'Anna, C.L. 1984. Chlorococcales (Chlorophyta) from the State of Sao Paulo, Brazil. J. Cramer, Vaduz, 348 p.

Sant'Anna, C.L. & Bicudo, C.E.M. 1981. Nomenclatural problems in *Scenedesmus bijugus* (Chlorophyceae/Scenedesmaceae). Taxon 30: 645-646.

Sant'Anna, C.L. & Martins, D.V. 1982. Chlorococcales (Chlorophyceae) from lakes Cristalino and Sao Sebastiao, Amazonas, Brazil: taxonomy and limnological aspects. Revista Brasileirade Botánica 5: 67-82.

Sant'Anna, C.L., Xavier, M.B. & Sormus, L. 1988. Qualitative study of the phytoplankton of the Serraria Reservoir, State of Sao Paulo, Brazil. Revista Brasileira de Biologia 48: 83102.

Sant'Anna, C.L., Azevedo, M.T.P. & Sormus, L. 1989. Phytoplankton of Lago das Garbas, Parque Estadual das Fontes do Ipiranga, Sao Paulo, SP, Brazil: taxonomic study and ecological aspects. Hoehnea 16: 89-131.

Schetty, S.P. 1998. Contribution to the planktonic phycoflora (except diatoms) of Lake Nymphaea, PEFI, Sao Paulo. Course Conclusion Monograph, Mackensie Presbyterian University, Sao Paulo, 97 p.

Schwarzbold, A. 1992. Effects of the flooding regime of the Mogi-Guacu River (SP) on the structure, diversity, production and stocks of the periphyton of Lagoa do Infernao. PhD thesis, Federal University of Sao Carlos, Sao Carlos, 154 p.

Smith, G.M. 1913. *Tetradesmus,* a new four-celled coenobic Alga. Bulletin of the Torrey Botanical Club 40: 75-87.

Smith, G.M. 1916. A monograph of the algal genus *Scenedesmus* based upon culture studies. Transactions of the Wisconsin Academy of Sciences, Arts and Letters 18: 422-530.

Smith, G.M. 1920. Phytoplankton of the island lakes of Wisconsin, 1: Myxophyceae, Phaeophyceae, Heterokonteae, and Chlorophyceae exclusive of the Desmidiaceae. Bulletin of the Wisconsin Geological and Natural History Survey 57: 1-243.

Smith, G.M. 1926. The plankton algae of the Okojobi region. Transactions of the American Microscopical Society 45: 156-233.

Silva, L.H.S. 1999. Phytoplankton of a eutrophic reservoir (Monte Alegre Lake). Revista Brasileirade Biologia 59: 281-303.

Souza, R.C.R. 2000. Spatio-temporal dynamics of the phytoplankton community of a hypereutrophic reservoir Salto Grande, Americana, Sao Paulo. PhD Thesis, University of Sao Paulo, Sao Carlos, 256 p.

Teiling, E. 1916. *Tetrallantos,* eine neue Gattung der Protococcoideen. Svensk botaniska Tidskrift 10: 59-65.

Toledo, L. & Comas, A. 1988. On the morphological variability and taxonomy of some species of Scenedesmus (Chlorellales). Acta Botanica Cubana 57: 1-32.

Trainor, F.R. 1963. Zoospores in *Scenedesmus obliquus.* Science 142: 1673-1674.

Trainor, F.R. 1965. Motility in *Scenedesmus* incubated in nature. Bulletin of the Torrey Botanical Club 92: 329-332.

Trainor, F.R., Rowland, H.L., Lylis, J.C. Winter, P.A. & Bonanomi, P.L. 1971. Some examples of polymosphism in algae. Phycologia 10:113-119.

Tsarenko, P.M. & Petlevanny, O.A. 2001. Doplolneniek "Raznoobraziju vodoroslej Ukrainy. Algologia, Suppl. p. 1-130.

Tucci, A. 2002. Phytoplankton community succession in an urban eutrophic reservoir, Sao Paulo, SP, Brazil. PhD Thesis, Universidade Estadual Paulista, Rio Claro, 274 p.

Tucci, A., Sant'Anna, C.L., Gentil, R.C. & Azevedo, M.T.P. 2006. Phytoplankton of Lago das Garcas, Sao Paulo, Brazil: a eutrophic urban reservoir. Hoehnea 33:147-175.

Tundisi, J.G. & Hino, K. 1981. List of species and growth seasons of phytoplankton from Lobo (Broa) reservoir. Revista Brasileira de Biologia41: 63-68.

Uherkovich, G. 1966. Die *Scenedesmus-Arten* Ungarns. Akademiai Kiadó, Budapest, 173 p.

Uherkovich, G. 1968. Zur Chlorococcalen Flora Finnlands, 1: Elenas-Tvarminne-Gegend. Acta BotanicaFennica 82: 1-26.

Vercellino, I.S. 2001. Succession of the periphytic algal community in two reservoirs of the Parque Estadual das Fontes do Ipiranga, São Paulo: influence of trophic state and climatological period. Master's dissertation, Universidade Estadual Paulista, Rio Claro, 176 p.

Van-den-Hoek, C., Mann, D.G. & Jahns, H.M. 1997. Algae: an introduction to phycology. Cambridge University Press, Cambridge, 627 p.

Watanabe, S. & Floyd, G. L. 1992. Comparative ultrastructure of zoospores with parallel basal bodies from the green algae *Dictyochloris fragrans* and *Bracteococcus* sp. American Journal of Botany 79: 551-555.

West, G.S. & Fritsch, F.E. 1927. A treatise on the British freshwater algae. Cambridge University Press, Cambridge, 534 p.

West, W. & West, G.S. 1902. A contribution to the freshwater algae of Ceylon. Transactions of the Linnean Society of London, Botánica series 6: 123-215.

Wille, N. 1897. Conjugatae and Chlorophyceae. *In:* Engler, A. & Prantl, K. (eds). Die natürlichen Pflanzenfamilien. Verlag von Wilhelm Engelmann, Leipzig, vol. 1, p. 1-161.

Wille, N. 1909. Conjugatae and Chlorophyceae. In: Engler, A. & Prantl, K. (eds). Die natürlichen Pflanzenfamilien. Verlag von Wilhelm Engelmann, Leipzig, vol. 1, p. 1-136 (2ª edition).

Wittrock, V.B. & Nordstedt, C.F.O. 1880. Algae aquae dulcis exsiccatae praecipue scandinavicae quas adjectis algis marinis chlorophyllaceis et phycochromaceis. O.L. Svanbäcks Boktryckeri Aktiebolac, Lundae, Fascículo 8.

Xavier, M.B. 1979. Contribution to the study of the seasonal variation of phytoplankton in Billings Reservoir, Sao Paulo. Master's dissertation, University of Sao Paulo, Sao Paulo, 146 p.

Xavier, M.B., Monteiro-Jr., A.J. & Fujiara, L.P. 1985. Limnology of reservoirs in the southeast of the state of São Paulo. Boletim do Instituto de Pesca 12: 145-186.

(SP96850): **Municipality of SAO PAULO**, Horto Florestal, lake, *C.E.M. Bicudo & R.M.T. Bicudo,* 16-VII-1962.

(SP96890): **Municipality of UBATUBA**, without precise indication of the place, *O. Montes & R.R. Martins,* 29-I-1966.

(SP96909): **Municipality of SANTO ANDRÉ,** Biological Station of Alto da Serra de Paranapiacaba, lake near the headquarters, *C.E.M. Bicudo,* 20-I-1966.

(SP96922): **Municipality of SANTO ANDRÉ,** Biological Station of Alto da Serra de Paranapiacaba, stream near the headquarters, *C.E.M. Bicudo,* 20-I-1966.

(SP96949): **Municipality of PINDAMONHANGABA,** BR-116, km 270-271, lagoon, *C.E.M. Bicudo,* 21-V-1966.

(SP96950): **Municipality of PINDAMONHANGABA,** Sao Joao farm, Sao Joao lake. Joao, *C.E.M.Bicudo,* 21-V-1966.

(SP96956): **Municipality of GUARATINGUETÁ,** BR-116, lake near the Clube dos 500, *C.E.M. Bicudo,* 01-IV-1966.

(SP96958): **Municipality of GUARATINGUETÁ,** BR-116, lake near Clube dos 500, *C.E.M. Bicudo,* 01-IV-1966.

(SP96965): **Municipality of GUARATINGUETÁ,** Clube dos 500, lake, *C.E.M. Bicudo,* 01-IV-1966.

(SP104098): **Municipality of SAO PAULO**, Santo Amaro, pond near Av. Washington Luiz n⁰ 3669, *B. Skvortzov,* 01-VI-1967.

(SP104430): **Municipality of ATIBAIA,** SP-381, km 527, lake, *L. Sormus,* 01-V-1973.

(SP104483): **Municipality of ATIBAIA,** SP-381, km 527, backwater next to a waterfall, *L. Sormus,* 01-V-1973.

(SP104685): **Municipio de SAO CARLOS,** km 234, Abatedouro de Aves Ito, lagoa de estabilizacao, *C.E.M. Bicudo, C.R. Leite, L. Sormus &P.A.C. Senna,* 10-V-1973.

(SP104723): **Municipality of SAO CARLOS,** SP-310, km 222, Aldeia Conde do Pinhal, stream, *C.E.M. Bicudo &L. Sormus,* 10-V-1973.

(SP104728): **Municipality of RIO CLARO,** SP-310, km 156, flooded, *C.E.M. Bicudo & P.A.C. Senna,* 10-V-1973.

(SP113429): **Municipality of IBIRÁ,** Termas de Ibirá, lagoon, *D.M. Vital,* 25-V-1973.

(SP113484): **Municipality of IRAPUA,** Novo Horizonte-Irapua highway, ca. 1 km from Cervo Grande stream bridge, flooded, *D.M. Vital,* 24-V-1973.

(SP113524): **Municipality of BRAGAN^A PAULISTA,** 3 km northeast of the city cf Braganca Paulista, pond, *D.M. Vital,* 20-VI-1973.

(SP113574): **Municipality of TAMBAÚ,** Clube de Tambaú, dam, *D.M. Vital,* 23-VI- 1973.

(SP113662): **Municipio de MOGI DAS CRUZES,** SP-88, km 74-75, lagoon, *C.E.M. Bicudo, C.R. Leite & L. Sormus,* 18-VI-1973.

(SP113664): **Municipality of JUQUIÁ,** BR-116, km 165, lake, *C.E.M. Bicudo & L. Sormus,* 01-III-1973.

(SP113669): **Municipio de REGISTRO,** Estagao Experimental do Instituto Agronómico de Campinas, lake, *C.E.M. Bicudo, C.R. Leite & L. Sormus,* 01-III-1973.

(SP113672): **Municipality of JUQUIÁ,** BR-116, km 160, flooded, *C.E.M. Bicudo & L. Sormus,* 01-III-1973.

(SP113673): **Municipality of MIRACATU**, Jaragatiá, Pettena farm, lake, *C.E.M. Bicudo, C.R. Leite & L. Sormus,* 01-III-1973.

(SP113679): **Municipality of MIRACATU**, Jaragatiá, Pettena farm, lake, *C.E.M. Bicudo, C.R. Leite & L. Sormus,* 01-III-1973.

(SP114511): **Municipality of SANTO ANASTÁCIO**, highway Santo Anastácio-Mirante do Paranapanema, river Santo Anastácio, *D.M. Vital,* 23-VIII-1973.

(SP114513): **Municipality of RANCHARIA**, Vila Agisse, Ribeirao Capivari, *D.M. Vital,* 25-VIII-1973.

(SP114515): **Municipality of IBIÚNA**, highway Cotia-Ibiúna, km 54, river Soroca-Mirim, *L. Sormus,* 19-IX-1973.

(SP114539): **Municipality of MOJI GUA£U**, no precise location, *D.M. Vital,* 17-X-1973.

(SP114545): **Municipality of DOIS CÓRREGOS**, without precise indication of the place, *D.M. Vital,* 19-X-1973.

(SP114558): **Municipality of PONTES GESTAL**, Turvo river, *D.M. Vital,* 23-XI-1973.

(SP123852): **Municipality of SUMARÉ**, Baptist Camp, lake, *L. Sormus,* 06-I- 1975.

(SP123858): **Municipality of ITIRAPINA**, Forestry Institute, stabilization pond, *O. A. da Silva,* 26-I-1975.

(SP123865): **Municipality of SUMARÉ**, Baptist Camp, lake, *L. Sormus,* 11 -II- 1975.

(SP123867): **Municipality of RIO CLARO**, Horto Florestal, lake, *O.A. da Silva,* 31-I- 1975.

(SP123886): **Municipality of SALESÓPOLIS**, Boracéia, COMASP Reserve, *P.A.C. Sema,* 12-X-1973.

(SP123900): **Municipality of PIRASSUNUNGA,** SP-225, km 23, Cascalbo neighborhood, lake, *P.A.C. Senna,* ?-XII-1973.

(SP123906): **Municipality of PIRASSUNUNGA,** SP-225, km 38, Air Force Academy, *P.A.C. Senna,* ?-XII-1973.

(SP130425): **Municipality of SOROCABA**, Brigadeiro Tobias, lake, *D.M. Vital,* 19-V- 1972.

(SP130427): **Municipality of RIBEIRÂO BRANCO**, without precise indication of location, *D.M. Vital,* 19-V-1972.

(SP130432): **Municipality of SÂO BERNARDO DO CAMPO,** Billings dam, Grande creek, *C.R. Leite,* 05-X-1972.

(SP130438): **Municipality of SÂO PAULO,** Cabuçu reservoir, R. *Roque,* 04-XIII-1972.

(SP130445): **Municipality of CAMPOS DE JORDÂO,** Alagoinha, lake, *M.M. Sakane,* 11-III-1973.

(SP130453): **Municipality of SÂO PAULO,** Parque Estadual das Fontes do Ipiranga, artificial lake, *C.R. Leite,* 05-VIII-1973.

(SP130784): **Municipality of AMERICANA,** Americana Dam, *G.Y. Shimizu,* 05- VII-1973.

(SP130785): **Municipality of BARRA BONITA**, Barra Bonita Reservoir, R. *Roque,* 27-VI-1973.

(SP130786): **Municipality of SÂO PAULO,** SABESP aerobic lagoon, R. *Roque,* 02- VIII-1973.

(SP130789): **Municipality of TUPÂ,** without precise indication of location, *D.M. Vital,* 20-VII- 1973.

(SP130790): **Municipality of BAURU,** Batalha river, near the Bauru-Piratininga road, *D.M.* Vital, 21-VIII-1973.

(SP130795): **Municipio de ITAPEVA,** rio Apiaí Mirim, *L. Sormus,* 21-IX-1973.

(SP130801): **Municipality of UBATUBA,** Caraguatatuba-Ubatuba road, km 220, pond, *C.R. Leite,* 22-X-1974.

(SP130804): **Municipality of UBATUBA,** swamp forest, *C.R. Leite,* 22-X-1974.

(SP130806): **Municipality of ITIRAPINA,** Forestry Institute, dam, *O.A. da Silva,* 261-1975.

(SP130808): **Municipality of CANANÉIA,** 2 km from ferry port, drain, *C.R. Leite,* 05- III-1975.

(SP130813): **Municipality of CANANÉIA,** Ilha Comprida, 120 m from the sea, lagoon, coll. *D.M. Vital,* 07-III-1975.

(SP130815): **Municipality of ARUJÁ,** Fiscal Club of Brazil, lake, *L. Sormus,* 01-V- 1975.

(SP130956): **Municipality of AVARÉ,** SP-310, km 276, lake, *O. Yano & R.C.A. Souza,* 23-I-1976.

(SP131583): **Municipality of SAO PAULO,** University City, Institute of Biosciences, artificial lake, *L. Sormus,* 24-II-1977.

(SP139733): **Municipality of ITU,** SP-280, km 77, lake, *C.R Leite,* 11-V-1977.

(SP139736): **Municipality of SOROCABA**, SP-280, km 84, lake, *C.R. Leite,* 11-V-1977.

(SP139741): **Municipality of PORANGABA**, SP-280, km 127, lake, *C.R. Leite,* 11-V- 1977.

(SP139745): **Municipality of BOFETE**, SP-280, km 197, lake, *C.R. Leite,* 11-V-1977.

(SP139746): **Municipality of ITATINGA**, SP-280, km 216, lake, *C.R. Leite,* 11-V-1977.

(SP139747): **Municipality of AVAÍ,** EstanciaAruana, lake, *C.R. Leite,* 12-V-1977.

(SP139749): **Municipality of AVAÍ,** EstanciaAruana, empopado, *C.R. Leite,* 12-V-1977.

(SP139750): **Municipality of PIRATININGA,** Água da Faca river, *C.R. Leite,* 12-V-1977.

(SP139752): **Municipality of PIRATININGA,** Batalha River, *O. Yano,* 12-V-1977.

(SP188210): **Municipality of SAO JOSÉ DOS CAMPOS,** SP-99, km 8, Vila Sao Judas, on the left, direction Sao José dos Campos-Caraguatatuba, pond, with reeds, *A.A.J. de Castro & C.E.M. Bicudo,* 21-II-1989.

(SP188211): **Municipality of MOJI DAS CRUZES,** SP-88, 1 km before Moji das Cruzes, direction

Salesópolis-Moji das Cruzes, *A.A.J. de Castro & C.E.M. Bicudo,* 21-II- 1989.
(SP188214): **Municipality of SAO MIGUEL ARCANJO,** SP-250, left side, 200 m towards Sao Miguel-Itapetininga, dam formed by the Acude stream, with water hyacinth, *A.A.J. de Castro & C.E.M. Bicudo &D.C. Bicudo,* 17-IV-1989.
(SP188215): **Município** *de* **ANGATUBA,** SP-270, km 203,7, on the right, direction Angatuba-Itapetininga, chácara Casa da Pedra, lake with Cyperaceae, *A.A.J. de Castro & C.E.M. Bicudo,* 17-IV-1989.
(SP188321): **Municipality of CASA BRANCA,** SP-340, km 228.5, stream, swamp effluent, right side, *A.A.J de Castro & C.E.M. Bicudo,* 17-X-1989.
(SP188322): **Municipality of SAO JOSÉ DO BARREIRO,** SP-64, km 0.8, Queluz-Sao José do Barreiro road, marsh, with *Thypha* and Cyperaceae, *A.A.J. de Castro & C.E.M. Bicudo,* 21-XI-1989.
(SP188323): **Municipality of SAO LUÍS DO PARAITINGA,** SP-125, km 34.7, right side, direction Taubaté-Sao Luís do Paraitinga, drain behind the school, amidst Cyperaceae, *Typha* and Liliaceae, *A.A.J. de Castro & C.E.M. Bicudo & E.M. De- Lamonica-Freire,* 27-XI-1989.
(SP188431): **Municipality of PILAR DO SUL,** SP-250, km 127, on the right, direction Sao Paulo-Pilar do Sul, Turvinho neighborhood, Turvinho river, periphyton among grasses, *A.A.J. de Castro, C.E.M. Bicudo&D.C. Bicudo,* 17-IV-1989.
(SP188434): **Municipality of ITANHAÉM,** SP-55, km 332.7, pond, with *Typha* and *Eichhornia, L.H.Z. Branco,* 28-II-1990.
(SP239042): **Municipality of ITAPETININGA,** SP-270/127, km 171, on the right, lake forming marsh, with Gramineae, *A.A.Jde Castro & C.E.M. Bicudo,* 11-IX-1990.
(SP239044): **Município de CAPIVARI,** SP-308, km 132, pond, with *Typha, Eichhornia* and *Pistia, A.A.Jde Castro & C.E.M. Bicudo,* 20-III-1990.
(SP239085): **Municipality of PARAGLACI' PAULISTA,** SP-284, km 457, stream after dam, no vegetation on banks or aquatic, *M.C. Bittencourt-Oltveira,* 20-VII-1991.
(SP239086): **Municipality of MARÍLIA,** SP-333, Água da Cobra stream, tributary of the Rio do Peixe, with ponds nearby, with some aquatic vegetation, *M.C. Bittencourt- Oliveira,* 20-VII-1991.
(SP239089): **Municipality of ASSIS,** SP-333, km 435, pond, with aquatic vegetation and tobacco on the banks,*M.C. Bittencourt-Oliveira,* 21-VII-1991.
(SP239091): Municipality of **INÚBIA PAULISTA,** SP-294, km 578, stream with vegetation almost completely covering it, M.C. *Bittencourt-Oliveira,* 20-VII-1991 (periphyton and phytoplankton).
(SP239092): **Municipality of DRACENA,** SP-294, km 644, near sand port, well altered stream, no aquatic vegetation, *M.C. Bittencourt-Oliveira,* 20-VII-1991.
(SP239095): **Municipality of PIRACICABA,** Piracicaba river, last waterfall downstream of the river, in front of the brewery, *A.A.J. de Castro,* 29-X-1991.
(SP239098): **Municipality of BRODOWSKI,** vicinal highway, km 7, on the left, direction Brodowski-Jardinópolis, in front of km 7, marsh with Cyperaceae and *Typha, A.A.J. de Castro,* 16-XI-1991.
(SP239136): **Municipality of TEODORO SAMPAIO,** Inhana stream, *M.C. Bittencort- Oliveira,* 08-XII-1991.
(SP239138): Municipality of **VARGEM GRANDE PAULISTA,** SP-270, km 42.4, on the right, direction Cotia-Vargem Grande, Chácara "Ise", dammed stream with masses of Cyanophyceae, on macrophytes, *A.A.J. de Castro,* 18-II-1992 (periphyton and phytoplankton).
(SP239140): **Município de PIRAJU,** Represa de Jurumirim, 300 m from the island, downstream of the dam, coll. *D.C. Bicudo & D.M. de Figueiredo,* 21-I-1992.
(SP239233): **Municipality of MONTE ALTO,** highway between Monte Alto and Vista Alegre, lake, with grasses and *Typha,* coll. *L.H.Z. Branco,* 20-II-1992.
(SP239236): **Municipality of LENCOIS PAULISTA,** SP-300, km 299.5, lake at entrance to town, Lengóis river, *C.M. Bicudo &D.C. Bicudo,* 22-II-1992.
(SP239237): **Municipality of URANIA,** SP-300, 1 km before the city, unspecified location, with Cyperaceae, grasses and *Myriophyllum, L.H.Z. Branco,* 05-XII-1991.
(SP239239): **Municipality of ARA^ATUBA,** Marechal Rondon highway, unspecified location, with Cyperaceae, grasses and *Myriophyllum, L.H.Z. Branco,* 15-I-1992.
(SP239241): **Municipality of GENERAL SALGADO,** highway Jesulino da Costa Frota (side road), 1.5 km from highway SP-310, unspecified location, with Cyperaceae, grasses and *Typha, L.H.Z. Branco,* 05-XII-1991.
(SP239242): **Municipality of MAIRIPORA,** Santa Inés road, Mairipora Dam, *M.C. Bittencourt-Oliveira,* 12-IV-1992.
(SP255722): **Municipality of MATAO,** SP-310, km 309, marsh, with *Typha* and *Eichhornia, L.H.Z. Branco,* 28-II-1990.
(SP255724): Município *de* **SAO PEDRO,** SP-304, km 127, lago do restaurante do lago, with *Nymphaea elegans* and *Salvinia,* perifiton, *A.A.J. de Castro & C.E.M. Bicudo,* 20-III- 1990.
(SP255728): Municipality of **MIRASSOL,** SP-31, km 410.7, pond with grasses, clay bottom, periphyton, *A.A.J. de Castro & C.E.M. Bicudo,* 10-IV-1990.
(SP255729): Município de **UCHOA,** SP-310, km 410.7, 30 km before Sao José do Rio Preto, pond with

Poaceae, clay bottom, periphyton, *D.C. Bicudo & C.E.M. Bicudo*, 10-IV-1990.

(SP255731): Municipality of **TREMEMBÉ**, SP-442, 13.2 km before Taubaté, lake on the right, direction Pindamonhangaba-Taubaté, with *Utricularia, Typha* and Cyperaceae, periphyton, *A.A.J. de Castro & C.E.M. Bicudo*, 24-IV-1990.

(SP255738): **Municipality of PEDREGULHO**, "Sobrado" farm, stream, stone scraping, sediment, tree trunk, phytoplankton, *A.A.J. de Castro*, 02-IX-1990.

(SP255739): **Municipality of GUARÁ**, District of Pioneiros, SP-330, km 393.25, highway between Guará and Sao Joaquim da Barra, near Sao Joaquim da Barra, on the right, direction, Guará-Sao Joaquim, pond with Cyperaceae and Poacae, *A.A.J. de Castro*, 02-IX- 1990.

(SP255757): Municipality of **PRESIDENTE VENCESLAU**, SP-563, km ?, marsh, with aquatic vegetation, clear water, periphyton,*M.C. Bittencourt-Oliveira*, 21-VII-1991.

(SP255760): **Municipio de ITAÍ**, SP-255, km 308.3, Represa de Jurumirim, right side, direction Itai-Taquarituba, presence of *Myriophyllum* and *Eichhornia*, perifiton, *A.A.J. de Castro, C.E.M. Bicudo &M.R.Marques-Lopes*, 10-IX-1991.

(SP255761): **Municipality of BATATAIS**, SP-330, km 355.5, right-hand side, direction Batatais-Franca, dam with little macrophyte, *Hydrocotile* and *Myriophyllum*, in front of the "Aparecida" seedling nursery, periphyton, *A.A.J. de Castro*, 16-XI-1991.

(SP255765): **Municipio de ELDORADO**, Ribeirao das Ostras, 300 m before Caverna do Diabo, rock scrape, periphyton, *C.E.M. Bicudo & D.C. Bicudo*, 29-XII-1991.

(SP255766): **Municipality of PIEDADE**, SP-232, km 132.3, dammed stream, 12.8 km before the Piedade-Ibiúna junction, periphyton, *C.E.M. Bicudo & D.C. Bicudo*, 30- XII-1991.

(SP255768): Boundary of the municipalities of **JAÚ and BARIRI**, SP-304, km 317.5, 13 km before Bariri, farm "Santa Fé", pond with aquatic plants, periphyton, *C.E.M. Bicudo & D.C. Bicudo*, 22-II-1992.

(SP255769): **Municipality of REGINÓPOLIS**, SP-331, km 115.2, on the left, towards Pirajui, right side of the Batalha river, 500 m after the entrance to Reginópolis, marsh with macrophytes, periphyton, *C.E.M. Bicudo &D.C. Bicudo*, 22-II-1992.

(SP255771): **Municipality of ITAJU**, SP-304, km 347.5, direction Ibitinga, acude, with aquatic plants, periphyton, *C.E.M. Bicudo &D.C. Bicudo*, 22-II-1992.

(SP255772): **Municipality of BARRETOS**, Barretos, in the city, lake region, with grasses and Cyperaceae, *L.H.Z. Branco*, 28-II-1990.

(SP336342): **Municipality of RIFAINA**, Rio Grande bridge connecting the municipalities of Rifainae Araxá, Phytoplankton, *C.E.M. Bicudo &D.C. Bicudo*, 30-V-2000.

(SP336346): **Municipio de JACUPIRANGA**, SP-139, km 23, pond, with cattails, coll. *C.E.M. Bicudo, L.A. Carneiro & S.M.M. Faustino*, 13-IX-2000.

(SP336348): **Municipio de CERQUEIRA CÉSAR**, SP-270, km 13, stream with strong current, phytoplankton, coll. *L.L. Morandi & S.P. Schetty*, 21-IX-2000.

(SP336349): **Municipio de NOVO HORIZONTE**, SP-304, km 451, pond, periphyton collected with net, col. *C.E.M. Bicudo, S.M.M. Faustino &L.R. Godinho*, 14-II-2001.

(SP355356): **Municipality of CACONDE**, SP-344, km 291, *Caconde* Reservoir, periphyton, *C.E.M. Bicudo, L.A. Carneiro & S.M.M. Faustino*, 08-VIII-2000. 21°34'39,9"S, 46°37'31,0"W, conductivity 30 pS cm⁻¹ , pH 8,0.

(SP355358): **Municipality of ITAPORANGA**, SP-255, km 358, river, periphyton, *S.MM. Faustino, &S.P. Schetty*, 26-VII-2000. 23°42'24.3"S, 49°28'15.6'W, conductivity 20 pS cm⁻¹ , pH 7.1.

(SP355360): Municipality of **PIQUETE**, road linking Lorena to Piquete, km 65, stream, periphyton, *C.E.M. Bicudo, D.L. Costa & F.C. Pereira*, 19-IX-2001. 22°37'24.2"S, 45°09'40.1"W. pH 6.2.

(SP355363): Municipality of **SAO LUIZ DO PARAITINGA**, SP-125, km 76, pond, periphyton, *C.E.M. Bicudo, D.L. Costa & F.C. Pereira*, 19-IX-2001. 23°21'58.8"S, 45°08'30.8"W, pH 6.0.

(SP355366): **Municipio de MACEDONIA**, rodovia Alberto Faria, sentido MiaEstrela- Macedonia, 2 km before the entrance to Macedonia, acude, perifiton, *C.E.M. Bicudo, D.L. Costa & S.M.M. Faustino*, 25-IV-2001. 20°08'19.5"S, 50°11'56.4"W, conductivity 70 pS cm⁻¹ , pH 6.6.

(SP355370): **Municipality of SANTA CRUZ DO RIO PARDO**, SP-225, km 309, acude, benthos, *C.E.M. Bicudo, L.A. Carneiro & S.M.M. Faustino*, 27-III-2001. 22°45'24.8"S, 49°29'07.7"W, conductivity 110 pS cm⁻¹ , pH 6.7.

(SP355374): **Municipality of MONTE APRAZÍVEL**, side road linking Monte Aprazivel to Engenheiro Balduino, 0.5 km, pond, periphyton, *C.E.M. Bicudo, D.L. Costa & S.M.M. Faustino*, 24-IV-2000. 20°45'18.0"S, 49°42'13.9"W. Conductivity 160 pS cm⁻¹ , pH 7.0.

(SP355377): **Municipio de LINS**, SP-300, km 436.5, marsh, periphyton, *C.E.M. Bicudo, L.R Godinho & C.I. Santos*, 14-VIII-2001. 21°43'53.2"S, 49°42'31.9"W, pH 6.3.

(SP355380): **Municipality of ORLANDIA**, highway Morro agudo towards Sao Joaquim da Barra, km 13, dam, *C.E.M. Bicudo & D.C. Bicudo*, 29-V-2000.

(SP355382): **Municipio de PITANGUEIRAS**, SP-322, km 368, acude, periphyton, *C.E.M. Bicudo, S.M.M. Faustino, & L.L. Morandi*, 16-VII-2000. 20°59'30,5"S, 48°14'01,1"W, conductivity 40 pS cm⁻¹ , pH 6,5.

(SP355385): Municipality of **SANTA ALBERTINA**, vicinal Vereador ítalo Biani, 15 km after the town, stream, benthos, *C.E.M. Bicudo, D.L. Costa & S.M.M. Faustino,* 24-IV- 2001. 20°3'20.1"S, 50°46'0.1"W, conductivity 110 pS cm^{-1} , pH 7.7.

(SP355386): **Municipality of SANTO ANTONIO DO ARACANGUÁ,** SP-320, km 463, lake, metaphyton, *C.E.M. Bicudo, D.L. Costa & S.M.M. Faustino,* 25-IV-2001. 20°50'30.2"S, 50°27'14.4"W, conductivity 20 pS cm^{-1} , pH 8.0.

(SP355388): **Municipality of ITAPURA,** SP-595, km 21.5, Tiete river, metaphyton, *C.E.M. Bicudo &D.C. Bicudo,* 16-V-2001. 22°16'41.0"S, 51°48'16.5"W, conductivity 40 pS cm^{-1} , pH 6.0.

(SP355389): **Municipality of COSMORAMA,** SP-320, km 496, acude, metaphyton, *C.E.M. Bicudo, D.L. Costa & S.M.M. Faustino,* 24-IV-2001. 20°30'18.4"S, 49°46'14.4"W, conductivity 30 pS cm^{-1} , pH 6.4.

(SP355390): **Municipality of SAO PEDRO DO TURVO,** BR-153, 10 km from the municipal border, marsh, metaphyton, *C.E.M. Bicudo, L.A. Carneiro & S.M.M. Faustino,* 28- III-2001. 22°48'46.3"S, 49°47'24.8"W, conductivity 60 pS cm^{-1} , pH 6.2.

(SP355391): **Municipio de GUAPIARA,** SP-250, km 284, rio Sao José, periphyton, *C.E.M. Bicudo, L.A. Carneiro & S.M.M. Faustino,* 27-III-2001. 24°19'12.0"S, 48°37'1.7"W, conductivity 30 pS cm^{-1} , pH 6.9.

(SP355392): **Municipio de ITABERÁ,** SP-249, km 114, periphyton, *S.M.M. Faustino & S.P. Schetty,* 26-VII-2000. 23°51'11,3"S, 49°09'10,8"W, conductivity 10 pS cm^{-1} , pH 6,9.

(SP355393): **Municipio de IPORANGA,** rio Fria, vicinal Iporanga-Apiaí, 36 km before Apiaí, periphyton, *C.E.M. Bicudo, L.A. Carneiro, & S.M.M. Faustino,* 13-IX-2000.

(SP355394): **Municipality of SANTA RITA DO OESTE,** side road, 3 km after entering the town, stream, metaphyton, *C.E.M. Bicudo, D.L. Costa & S.M.M. Faustino,* 25- IV-2001. 20°07'36"S, 50°48'0.9"W, conductivity 110 pS cm^{-1} , pH 6.8.

(SP355396): **Municipality of MIRANTE DO PARANAPANEMA,** SP-272, km 30.5, acude, metaphyton, *C.E.M. Bicudo & D.C. Bicudo,* 16-V-2001. 22°16'41.0"S, 51°48'16.5"W, conductivity 40 pS cm^{-1} , pH 6.0.

(SP355398): **Municipality of TURMALINA,** SP-462, km 14, stream, metaphyton, *C.E.M. Bicudo, D.L. Costa & S.M.M. Faustino,* 25-IV-2001. 20°09'20.2"S, 50°26'16.2"W, conductivity 70 pS cm^{-1} , pH 6.6.

(SP365686): **Municipality of JUNDIAÍ,** SP-360, km 68, stream, *C.EM. Bicudo & S.P. Schetty,* 05-V-2000.

(SP365687): **Municipality of LIMEIRA,** SP-151, between km 3 and 4, agude, *C.E.M. Bicudo & S.P. Schetty,* 05-V-2000.

(SP365688): **Municipio de RIBEIRAO BONITO,** SP-215, km 171, Rio do Pantano, *C.E.M. Bicudo & L.L. Morandi,* 12-V-2000.

(SP365690): **Municipality of MIGUELÓPOLIS,** dam, *C.E.M. Bicudo &D.C. Bicudo,* 30-V-2000.

(SP365691): **Municipality of PEDRO DE TOLEDO,** highway Manoel da Nóbrega, km 370.5, agude, *C.E.M. Bicudo &S.M.M Faustino,* 11-VII-2000.

(SP365693): **Municipio de CAPAO BONITO,** SP-127, km 199.9, river, *C.E.M. Bicudo, F.C. Pereira&L.L Morandi,* 18-VII-2000.

(SP365696): **Municipality of FARTURA,** SP-287, direction Fartura-Taquarituba, km 3, stream, *S.M.M. Faustino & S.P. Schetty,* 26-VII-2000.

(SP365697): **Municipality of DIVINOLÁNDIA,** DVL-040, 13 km from the junction with highway SP-344, waterfall, *C.E.M. Bicudo, L.A. Carneiro & S.M.M. Faustino,* 08-VIII-2000.

(SP365698): **Municipio SAO JOSÉ DO RIO PARDO,** SP-350, km 265, agude, *C.E.M. Bicudo, L.A. Carneiro & S.M.M.Faustino,* 08-VIII-2000.

(SP365699): **Municipality of PIRACAIA,** SP-36, between km 101 and 102, Jacareí river reservoir, *C.E.M. Bicudo & C.I. Santos,* 24-IV-2000.

(SP365700): **Municipio de RINCAO,** SP-257, km 11, stream, *C.E.M. Bicudo, S.M.M. Faustino & L.L.Morandi,* 15-VIII-2000.

(SP365701): **Municipio de PRADÓPOLIS,** SP-291, Pradópolis, stream, *C.E.M. Bicudo, S.M.M. Faustino &L.L. Morandi,* 15-VIII-2000.

(SP399779): **Municipality of SAO PAULO,** Parque Estadual das Fontes do Ipiranga, Lago do IAG, periphyton, *I.S. Vercellino,* 03-VIII-1998. 23°38'08"S and 23°40'18"S, 46°36'48"We46°38'00"W.

(SP399780): **Municipality of SAO PAULO,** Parque Estadual das Fontes do Ipiranga, Lago do IAG, periphyton, *I.S. Vercellino,* 06-III-1999. 23°38'08"S and 23°40'18"S, 46°36'48"We46°38'00"W.

(SP399781): **Municipality of SAO PAULO,** Parque Estadual das Fontes do Ipiranga, Lago das Gargas, periphyton, *M. Borduqui,* 12-VII-2006. 23°38'08"S and 23°40'18"S, 46°36'48"We46°38'00"W.

(SP399782): **Municipality of SAO PAULO,** Parque Estadual das Fontes do Ipiranga, Lago das Gargas, periphyton, *M. Borduqui,* 17-I-2007. 23°38'08"S and 23°40'18"S, 46°36'48"We46°38'00"W.

(SP399783): **Municipality of SAO PAULO,** Parque Estadual das Fontes do Ipiranga, Lago das Ninféias, periphyton, *T. Santos,* 03-VIII-2007. 23°38'08"S and 23°40'18"S, 46°36'48"We46°38'00"W.

(SP399784): **Municipio de SÃO PAULO,** Parque Estadual das Fontes do Ipiranga, Lago das Ninféias, perifíton, *T. Santos,* 03-VIII-2007. 23°38'08"S and 23°40'18"S, 46°36'48"We46°38'00"W.